11-30-76

U.S. OBSERVATORIES:
A DIRECTORY AND TRAVEL GUIDE

H. T. Kirby-Smith

VNR VAN NOSTRAND REINHOLD COMPANY
New York Cincinnati Toronto London Melbourne

On the cover: Aerial view of Kitt Peak. In the foreground is the McMath Solar Telescope and a vacuum solar tower; on the highest point is the Mayall 158-inch telescope, with the University of Arizona's telescopes a little below. The view extends two hundred miles, across the U.S. Southwest and into Mexico. (Kitt Peak National Observatory Photograph)

Copyright © 1976 by Litton Educational Publishing, Inc.
Library of Congress Catalog Card Number 76-4448
ISBN 0-442-24451-7 (cloth)
ISBN 0-442-24450-9 (paper)

Published in 1976 by Van Nostrand Reinhold Company
A Division of Litton Educational Publishing, Inc.
450 West 33rd Street
New York, NY 10001

Van Nostrand Reinhold Limited
1410 Birchmount Road
Scarborough, Ontario M1P 2E7, Canada

Van Nostrand Reinhold Australia Pty. Ltd.
17 Queen Street
Mitcham, Victoria 3132, Australia

Van Nostrand Reinhold Company Ltd.
Molly Millars Lane
Wokingham, Berkshire, England

16 15 14 13 12 11 10 9 8 7 6 5 4 3 2 1

Library of Congress Cataloging in Publication Data

Kirby-Smith, Henry Tompkins, 1938–
    U. S. observatories.

    Bibliography: p.
    Includes index.
    1. Astronomical observatories—United States—
Directories.   2. Astronomical museums—United
States—Directories.   3. Planetaria—United States
—Directories.   I.   Title.
QB81.K57   522'.1'0973   76-4448
ISBN 0-442-24451-7
ISBN 0-442-24450-9 pbk.

CONTENTS

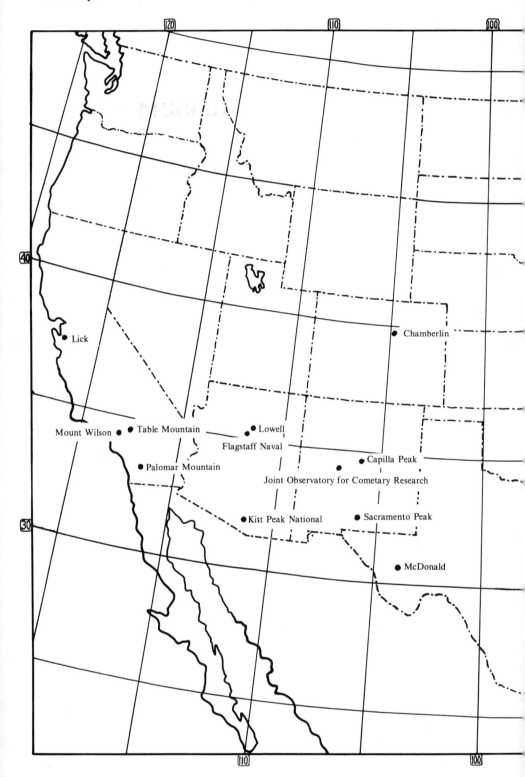

Lick

Chamberlin

Mount Wilson ● ● Table Mountain ● Lowell
Flagstaff Naval

● Palomar Mountain ● Capilla Peak
Joint Observatory for Cometary Research

● Kitt Peak National ● Sacramento Peak

● McDonald

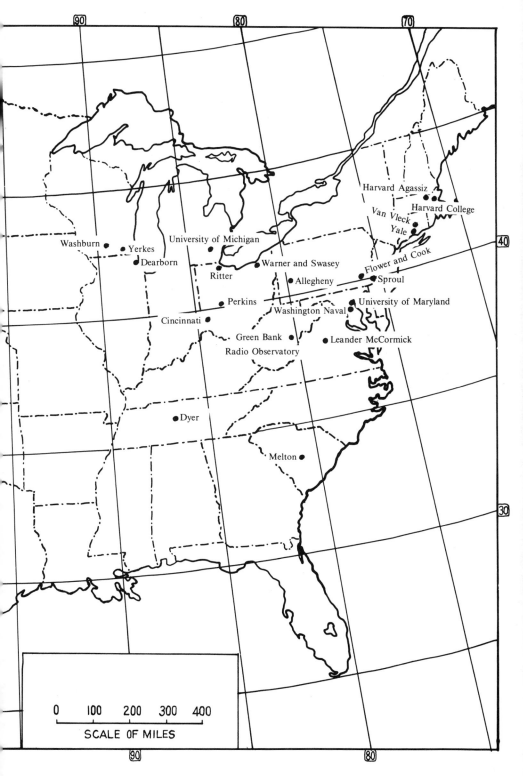

SCALE OF MILES

0   100   200   300   400

5

INTRODUCTION

The idea for this book germinated somewhere between the semiarid and the subhumid regions of Texas, during day-long drives across the entire state. My first plan was to provide a brief, informal guide to some of the major observatories that I had visited, almost by accident, in the course of the previous six months. "It ought to take about ten hours of work," I thought. It was only when I began to look for a standard list, to make sure that I did not leave out something of importance, that I became aware that there was no good list; so I set out to make one. More than three hundred observatories, and at least as many hours later, I began to realize that part of the problem was the question, "What is an observatory?" I am still uncertain about the answer.

At the very least, it must be something in existence. I am grateful to the Navy for their *Ephemeris* list, and for supplying me with a street address for one fugitive establishment, Mummy Mountain Observatory; but I cannot refrain from telling that Mummy Mountain's envelope was returned by the Postal Service marked "DECEASED." I have, however, attempted to include every viable observatory in the country. Some lively places have probably been overlooked, despite my best efforts; a few installations included may, like Mummy Mountain, belong among the living dead.

This book is intended to serve several purposes. Professional astronomers may find it interesting as a historical survey, and perhaps even helpful as a summary report of research (much of it elementary and routine, to be sure) under way in 1975. Should any project involving wide cooperation seem desirable (as in the "moonwatch" program), it may be useful to have some idea of the location of various kinds of equipment; perhaps the addresses alone might assist in organized projects to search for novas or other variable objects. Amateur astronomers may see the book as a travel guide, both to the great observatories that they may wish to visit and to many campgrounds, especially in the Southwest, where they can use their own equipment to best advantage. Teachers at all levels should, I hope, be able to use the catalog to plan field trips or tours. Perhaps the descriptions and photographs will also serve armchair travelers.

I have not attempted, in the lengthier descriptions of the great observatories, to provide a uniform presentation. Instead, I have emphasized a few things of particular interest that make each installation unique. Histories, biographies, monographs, and feature articles furnish more information than can be usefully digested in a few pages, and the *Bulletin of the American Astronomical Society*'s Observatory Reports cover technical innovations in detail. My treatment is variously historical, biographical, semi-technical, and geographical—and is at times a personal travelogue.

Advice for Astronomical Tourists

Prospective visitors should realize that no one—not even the most advanced amateur—should expect to look at or through an observatory's equipment except at the times and under the conditions set for visits. I write this as someone who at one time had pleasant illusions to the contrary. Time on the larger instruments is parceled out months or even years in advance; at small observatories, it is an added burden on a faculty member or a club officer to be on hand for public visits. Larger telescopes are in any case seldom used for visual work; usually they take photographs or operate with measuring equipment of various kinds. Actually, for visual observation of the moon and planets, a small instrument may be preferable; one contemporary expert on the planets, Gérard de Vaucouleurs, for example, used an 8-inch refractor for many of his fine drawings of Mars. Observatories are unheated and may be air-conditioned even on chilly days, in preparation for cooler night temperatures. (This prevents heat currents that make "seeing" bad.) Dress as you would for temperatures outdoors on that day. Children under five should not be taken to observatories; some places forbid children under ten or twelve. The reason is that—particularly in the dark—children start running around in circles, and the reverberation of shouts and squeals from circular walls makes them run faster. Even some planetariums do not admit small children. For observatory visits, large numbers of children should probably be separated into groups of five. Family groups are often welcome, however. Remember that unless a telescope is at least ten inches in aperture, and the skies are quite dark, a first view of an object such as a globular cluster (which looks splendid in a photograph) will be disappointing. As compensation, muggy, hazy, polluted city air may very well be quite steady, and provide fine views of the moon and the bright planets.

As a rule, the more an institution depends on public good will, the more accommodating it is to the public. There are notable exceptions to this: Lowell Observatory is independent and private—yet it treats visitors with great courtesy; at the other extreme are one or two state university observatories, which receive federal as well as state funds, that do their best to ignore the public. It is a pleasure, though, to have discovered that many great research centers—Kitt Peak, Allegheny, Harvard-Smithsonian, Leander McCormick, Green Bank, McDonald, Lowell, Lick, the Naval Observatories, Sacramento Peak, and others—do encourage the interested and intelligent visitor and suffer patiently the ignorant and the inane. Good scientists do not forget that a sense of mystery, or just a healthy curiosity, animates the most valuable endeavors.

There is also a great deal of hospitality among smaller observatories (though the expression "visits by arrangement" that occurs very frequently may mean a variety of things).

An excess of visitors is a bane to any busy teacher or research establishment, and one may sometimes sense this. But if you do happen to show up at a slack time, and express an informed interest in what you see, you may well get an hour or two of astronomical education that no amount of tuition money could buy.

Some Basic Information

Readers who are among the two-hundred-thousand persons who have taken an astronomy course recently should have few problems with this guide. Those who are beginners in astronomy should buy an introductory book; Bantam's "Knowledge Through Color" *Astronomy* is cheap and excellent. College bookstores will provide more advanced texts, some of which are mentioned at the back of this book. Visits to observatories are themselves opportunities to learn a lot.

Some terms used frequently in this book are:

Objective. The main lens or mirror of a telescope. The larger it is, the brighter the image is and the sharper it is (i.e., its *resolution* is better); magnification is not as important as most people think, but a large telescope objective does permit high magnification.

Ocular or **Eyepiece.** A combination of small lenses that sends light from the objective through the lens of the eye, where it is focused on the retina.

Refractor. A telescope with an objective lens, nearly always of two *elements* to correct color problems; the lens usually has a long *focal length*, the path that the light covers converging on a *focal point*. Refractors may have tubes 30 to 60 feet long. The best and largest of these were made between 1850 and 1910 in the United States. Color distortions increase with size, and some say that the glass sags too much to make a lens more than 40 inches in diameter (see Yerkes Observatory).

Reflector. A telescope in which the objective is a mirror, usually a *paraboloid*, curved to catch and focus the light. The main mirror can be supported in many places across its back or around its edges to prevent sagging; mirrors do not have *chromatic aberration* (false color). Large reflectors can be used at their *prime focus;* the mirror then acts as a giant camera and records star images directly on a photographic plate.

Cassegrain. A form of reflector with a hole bored in the center of the *primary*, the main mirror. A *secondary*, or small, convex mirror, reflects the light backwards through the hole. It is more compact and easier to attach instruments to than a *Newtonian*, which uses a diagonal mirror.

Schmidt. The most common kind of *catadioptric* telescope, one which uses both lenses and mirrors; somewhat like the *Maksutov*. The version for visual work is called the *Schmidt-Cassegrain*, and is rather common now in small observatories. The distinguishing character of the Schmidt is a thin lens at the front of the tube; this corrects some faults of reflectors, and makes possible a wider field. A *Schmidt Camera* has a photographic plate instead of a secondary mirror.

Coudé. A telescope in which two or more mirrors send the light path to the side of a reflector, and down the *equatorial shaft*. The focal path is made very long by using a secondary mirror even more convex than that for the Cassegrain focus. Usually this configuration is used for *spectroscopy* (see below).

Astro-Camera. In a sense, all telescopes are astro-cameras. The term usually means a combination of several refracting lenses that can photograph an area of the sky several degrees across. These instruments are much like larger versions of ordinary cameras.

Equatorial Mounting. Most telescopes are mounted so that one axis is parallel to the axis of the earth; this means that once they are aimed at a star, that axis then can be turned to compensate for the rotation of the earth. A *clock drive* turns the telescope so that it will follow what it is pointed at.

Photometry. The measurement of brightness of objects. This can be done visually, but with more precision by using a film (*photographic photometry*) or a cell or counting device, analagous to a camera's exposure meter (*photoelectric photometry*).

Spectroscopy. Prisms and *diffraction gratings* separate light into different wavelengths. *Spectrography* is the recording of these wavelengths. Such studies can involve the sun (*solar*), stars (*stellar*), or whole galaxies (*galactic*). This is the most valuable source of information about stars. Other terms to look up in an astronomy text on this subject are: *Doppler shift, emission lines, absorption lines,* and *red-shift.*

Astrometry. The measurement of star positions in the sky, in order to learn the motions, distances, and masses of stars (and, ultimately, of galaxies).

Radio, Infrared, Ultraviolet, X-ray, and **Gamma-ray Astronomy.** These forms of energy are like light in that they move at the same speed and are electromagnetic radiations. Radio and infrared have longer wavelengths (and lower frequency) than light; the

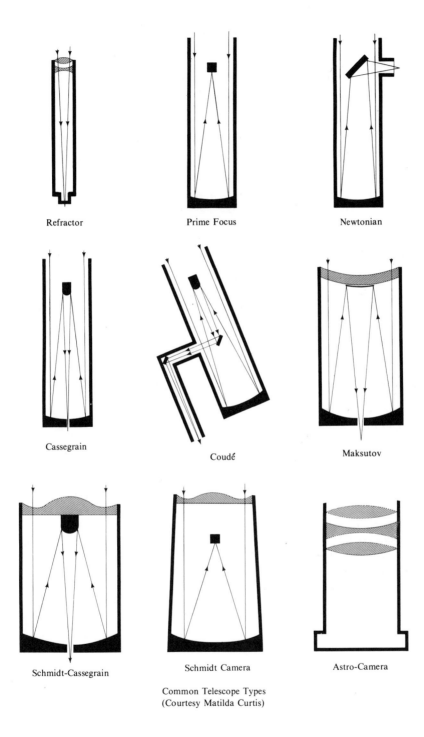

Refractor   Prime Focus   Newtonian

Cassegrain   Coudé   Maksutov

Schmidt-Cassegrain   Schmidt Camera   Astro-Camera

Common Telescope Types
(Courtesy Matilda Curtis)

others are higher-frequency and shorter wavelength. Radio waves must be gathered in big metallic dishes, or long arrays of TV-like antennas. Optical telescopes can focus some infrared and ultraviolet light, though these are invisible to the eye. Special films must be used to record them. Many devices are currently being invented to detect X rays and gamma rays; some of these must be carried above the earth's atmosphere in satellites. For detailed descriptions of the operation of radio telescopes see the discussions of Green Bank and of the Clark Lake Radio Observatory (in Part II: California).

Acknowledgments

I am grateful to the hundreds of persons who answered questionnaires and who provided material that supplied, confirmed, or supplemented information in this book. I wish that I could thank by name the observatory tour guides who imparted enthusiasm as well as facts; I did not know that I would write about what they told me, or I would have asked their names. I can mention Ted Rafferty of the U.S. Naval Observatory in Washington, though, and thank him. I am grateful to Deborah Jean Warner of the Smithsonian for the use to which I put her book on the Clarks, and for a letter bringing its telescope catalog up to date. Richard Y. Dow of the National Research Council sent a most encouraging and helpful letter just as I was getting under way with the book. The following persons sent me information and, on occasion, some kind notes: G. B. Field, director, and James C. Cornell of Harvard-Smithsonian; Paul B. Sebring, director of Haystack; John A. Kessler of M.I.T.; Joost Kiewiet de Jonge, acting director at Allegheny; Laurence W. Fredrick, director of McCormick; H. D. Craft, Jr., director of Arecibo; Harlan J. Smith, director of McDonald; Richard C. Altrock of Sacramento Peak; and H. W. Babcock, director of Hale Observatories. I caused Donald E. Osterbrock, Lick director, and Leonard V. Kuhi, chairman at Berkeley, more trouble than I intended. The librarians at the University of North Carolina at Greensboro are always helpful, especially Elizabeth Holder. I would not have undertaken this project without an unpaid leave of absence permitted me by UNC-Greensboro. I would like to thank William Lane, head of the English Department, for encouragement on this book and for permitting me to teach astronomy as part of my regular course load. I am especially grateful to E. E. Posey, head of the Mathematics Department, for having invited me to teach a section of astronomy some years ago. To the extent that a book such as this can be dedicated, it is dedicated to Professor Posey. I would like to indulge in the male chauvinist cliché of thanking my wife, who was most encouraging and helpful. Leslie Wenger, of Van Nostrand Reinhold, gave the manuscript a most meticulous (but not in the least picky) grooming. When a work such as this is first published it cannot escape errors and inconsistencies; the author must accept responsibility for these, but with a cheerful awareness that as his faults are disclosed his book will improve. In fact, the great majority of descriptions were submitted to individual observatories in draft form for correction. I acknowledge one more debt, to those who gave me their criticisms in advance.

PART I
FIFTEEN MAJOR OBSERVATORIES

The Center for Astrophysics:
Harvard College Observatory
and
Smithsonian Astrophysical Observatory
Cambridge, Massachusetts 02138

To put in conjunction the words Harvard and Smithsonian—as they were officially conjoined in 1973—is to designate not a place nor an assemblage of equipment, but rather a kind of astronomical personality. The two names belong together, certainly. To the extent that they have been connected with astronomy, they have meant flexibility, enterprise, international connections, an appreciation for the public interest in astronomy, a certain amount of crankiness, an indulgence towards amateurs, initiative in undertaking projects of doubtful outcome, appreciation of the personal history of scientists and the peculiar histories of scientific institutions, sophisticated naiveté, hospitality toward women scientists, and a genius for organization. In reading book-length histories of both institutions and in studying their current undertakings, one becomes aware that they display simultaneously the virtues of youth and old age, without the vices of either.

In view of the hundreds of universities in the United States that have telescopes of substantial aperture, and the hundreds of thousands of small telescopes in private hands, it is hard to believe that a century and a half ago there was not a single observatory in this country. The first building dedicated to such a purpose did not appear until 1831, on the campus of the University of North Carolina (see Morehead Planetarium). The Naval Observatory had no building to house its own meager instruments until 1844. When we learn, then, that it was not until 1847 that Harvard's "great refractor" of 15-inch aperture was finally in place, we must realize that this was still the most important first step for astronomical research in the western hemisphere.

As in the case of the Allegheny Observatory a few years later, a bright comet (in 1843) did the public-relations work. Before that, the Harvard Observatory consisted of a dome on top of Dana House and a few small telescopes that the first director, William Cranch Bond, found at Harvard or brought with him. These instruments were of almost no value in studying the comet; Bond lacked a micrometer to measure its diameter and graduated circles to help calculate its orbit. A public subscription raised $20,000 to buy the largest refractor available; on a challenge, by which David Sears pledged $5,000 for an observatory tower if the rest could be raised elsewhere, a total of eighty-two persons, seven businesses, and three nonprofit groups amassed this amount in six weeks. Those who doubt the influence of comets should consider this—and examine enrollment figures for astronomy classes before Comet Kohoutek did its vanishing act.

For much of the nineteenth century it was assumed that anything valuable in the way of scientific equipment would have to come from the other side of the Atlantic. Although this prejudice dogged the careers of both Alvan Clark and John Brashear, it was probably correct in 1843 to order the refractor lens from the German firm of Merz and Mahler, who were continuing the optical work of Fraunhofer, maker of the Dorpat Refractor. Harvard undertook to compete with none other than the Czar of Russia, demanding an

The 15½-inch "Great Refractor" made by Merz and Mahler in 1843. (Harvard College Observatory Photograph)

instrument equal to that at Pulkovo. The contract carefully specified that the company was to make two lenses, and that representatives of the Observatory were to be permitted to test them and choose the better. Considering the difficulty of communications, distances, delays, and the expense (perhaps a half-million dollars in today's terms), it was hardly paranoid of the two men who made the trip to Munich to insist that they be allowed to engrave identifying inscriptions with a diamond in the edge of the two lens elements. On rare occasions, when the lens is cleaned, the inscriptions proving that the correct lens was shipped can still be seen.

Shortly after the installation of the "great refractor," three other important instruments were added: a transit circle made by Troughton and Simms of London; a comet seeker, or short-focus, wide-angle telescope, made by Merz and Mahler; and a double-image micrometer.

The first important discovery of the Observatory came immediately, when W. C. Bond turned the 15-inch lens on Saturn and saw a shaded portion of the inner ring, which had not been noticed before. His son, using the other Merz and Mahler telescope, discovered eleven comets. With the development of photography, the "great refractor" became the first telescope to be used as an astro-camera.

The Sears Tower, with its 30-foot copper-covered wooden dome, is still in place in Cambridge; and the "great refractor," now used for student and visitor observation, is still there, its wood-and-paper tube with spherical weights projecting to the rear having furnished a model for the earlier products of the Alvan Clark enterprise.

Harvard College Observatory *circa* 1887. (Harvard College Observatory Photograph)

The year 1879 might be selected as the date when the Harvard Observatory settled down to its most serious work. With a new director (Edward C. Pickering, who took over in 1877) and a new piece of equipment, a careful photometric survey of the 4,000 brightest stars began. It was perhaps unfortunate that Polaris served as the comparison star, though the range of variation of this Cepheid did not vitiate these initial visual estimates. A few years later, the widow of Henry Draper—a physician and pioneer astrophotographer—made available both money and telescopic equipment that Pickering at once put to work. The project was enormous: to photograph and classify by spectrum some 10,000 stars. This became one of those labors—like Greek lexicons and the Shakespeare *Variorum*—that outlives the laborers. But the facts patiently accumulated and classified—by long-suffering men with cramped and chilled muscles by night and a team of dedicated women by day—laid the foundations for great structures of astrophysical theory. The constituents and temperatures of stars, spectroscopic binaries, the H-R diagram, and all the work in stellar evolution could not have been known or undertaken without this massive collocation.

In the 1890s, Pickering established a southern observing station in Arequipa, Peru, where photometric and spectral studies were extended to thousands of southern-hemisphere stars. The most remarkable outcome of photographic work here, though, occurred back in Cambridge in 1904, with the discovery of the period-luminosity relationship of Cepheid variables. It was Miss Henrietta Leavitt who, poring over plates made of the Magellanic clouds, found that the intrinsic brightness of these variables could be tied to the length of time that it took them to dim and brighten. This was perhaps the most valuable discovery ever made for determining distances within our own galaxy and beyond.

The 84-foot radio telescope at Agassiz Station, Harvard, Mass., operated jointly by HAO and SAO. (Harvard College Observatory Photograph)

By 1931 light and atmospheric pollution in Cambridge had made many projects in optical astronomy impossible there. A new observatory site, first called Oak Ridge and, since 1952, Agassiz Station, was acquired about 25 miles northwest of the University. This is a few miles beyond Walden Pond, just south of Route 2 at Harvard, Mass. Much equipment was moved here from Cambridge, and the fork-mounted 61-inch Wyeth reflector, made by J. W. Fecker, was built in 1932.

Over the years, the Harvard Observatory has acquired, put into use, transferred and retransferred, and installed in its southern-hemisphere stations, first in Peru and then in South Africa, an immense variety of equipment. This includes refractors still in use in Cambridge, a 24- by 33-inch Schmidt purchased for Agassiz Station, two dozen or more astrographic camera lenses ranging from a 1½-inch Cook to the 16-inch Metcalf doublet, some large reflectors, auxiliary equipment of all types and ages, and so on and so on. Presently there is an 84-foot radio antenna at Harvard, Mass., plus a station in Palestine, Texas.

At various times the Observatory has operated sky-survey and meteor-patrol cameras. As recently as 1965 a whole set of brand new cameras were put to work in this way, operating in gangs of three to photograph the sky simultaneously in blue, yellow, and red light.

Pickering originated the American Association of Variable Star Observers, and ties with this most productive amateur organization remained close during the directorship of Harlow Shapley. Although the university withdrew funding of the AAVSO after Shapley's death, that organization (now generously endowed by the Ford bequest) survives and continues an association with the Observatory.

There are also informal ties with the Sky Publishing Corporation, which puts out *Sky and Telescope* (the best magazine devoted to a single branch of learning), and which also prints books and pamphlets.

In July, 1973, the Center for Astrophysics was established in Cambridge, with a single director in charge of scientific activities of Harvard College Observatory and the Smithsonian Astrophysical Observatory. This act was simply a formal sign of the degree to which the two independent organizations had been cooperating for years; the SAO had actually relocated from Washington, D.C., to Cambridge in 1955, at which time Fred Whipple was named director while continuing to be on the staff of the Harvard Observatory. George B. Field, formerly chairman of the Department of Astronomy at Berkeley, has headed the Center since the 1973 reorganization.

The history of the SAO is far more embroiled with Washington politics and national economic vicissitudes than one could imagine. Aside from expressions of hope, little was accomplished in the way of setting up a national observatory in Washington until Samuel Pierpont Langley arrived on the scene from the Allegheny Observatory in 1887. He did not make much headway for a long time. An attempt to smuggle in an observatory by including it in plans for the zoo in Rock Creek Park encountered obstinate resistance. First, there were congressmen who thought that looking at monkeys was at best a waste of time and, at worst, likely to encourage a belief in evolution. Then the zoo was approved, but the proposed observatory eliminated. In the end, though, a $10,000 appropriation for astronomical work did get through the legislative session of 1892.

Actually work was already under way in a temporary wooden shed behind the main Smithsonian building, where the temperature could go as high as 120 degrees F. on a summer day. Langley had already invented the bolometer and observed lines in the "invisible" (infrared) part of the sun's spectrum. Because of this early specialization and its success—but also because the effect of sunlight on crops, weather conditions, and other practical things—a great deal of the Observatory's work was to involve setting up stations for solar observation. It was hard for a congressman to deny that the sun was important (though some did even that), and once the instruments and techniques were developed, they could with minimal cost be transferred to new stations for confirming measurements. In Smithsonian photographic archives there are scenes from California, Chile, and the Sinai that much resemble one another: a parched and rocky hillside with a walled observing cave dug into it, two or three astronomers, and a heliostat. Mountains were climbed, eclipses were chased, laborers hired, mules driven, delicate equipment packed and unpacked, baths taken in dishpans, and heat, cold, solitude, and boredom endured by parties of scientists and their wives. The results of fifty years of such expeditions include the determination of the solar constant of radiation, measurement of variation in solar radiation, effects of these things on the weather, atmospheric transmission of solar energy, determinations of the sun's temperature and its energy spectrum curve, and other radiometric and atmospheric studies. A corollary to this work was the development of many types of measuring devices.

Presently the SAO has a 60-inch reflector at its Mount Hopkins Station south of Tucson, Ariz.; under construction is the multiple-mirror telescope. Both are described in the Arizona section of Part II of this book.

At eleven locations around the world are the super-Schmidt Baker-Nunn tracking cameras, first astronomical tools of the space age. These are coupled with tracking lasers for following satellites, but they are also fine instruments for confirming reports of novas and comets.

At the end of August, 1975, the brightest nova in thirty years appeared in a position that made it especially conspicuous just northeast of Cygnus, the "northern cross."

An early solar research station of the Smithsonian Astrophysical Observatory, *circa* 1900, probably in California. (Photo: Smithsonian Institution)

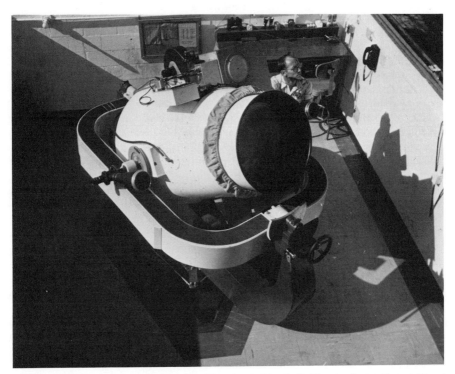

Three-ton Baker-Nunn camera with super-Schmidt optics used by the Smithsonian for tracking satellites at eleven sites around the world. (Smithsonian Astrophysical Observatory Photograph by Tom Butler)

Even after it had been reported in the newspapers, many an ignorant amateur, including this writer, continued to make "independent discoveries." The telephone number on which all seekers of instant celebrity pin their hopes is that of the Central Bureau for Astronomical Telegrams, whose headquarters have been at the Smithsonian offices in Cambridge since 1965. For many years before that, the Harvard Observatory had acted as a sort of western branch of this service, operated from Copenhagen after the First World War. Upon receipt and confirmation of a report on a comet or nova, the Bureau issues "astrograms," telegraphic messages to subscribing observatories around the world that alert them to the object. Printed circulars are mailed out as soon as possible to a wider group of subscribers. The Bureau also tries to adjudicate claims for the naming of comets, no more than three discoverers per comet being allowed by international convention. To conclude the story of Nova Cygni: within hours of the onset of nightfall (and a half-day after the actual discovery in Japan) the Bureau had received so many calls that it was forced to answer them with a tape recording. One's hopes of immortality dim when one hears a recorded voice that begins: "A bright nova has been discovered. . . ."

Another wide-ranging program that was only recently terminated was the "moon-watch" network established in the mid-fifties by Fred Whipple. The purpose of this was to track the early artificial satellites, using volunteer observers equipped with small, wide-angle telescopes.

In connection with Mr. Whipple, a brief story illustrates the tolerance of ignorant curiosity and the openness to amateur interest of these institutions. Ten years ago an especially bright meteor shower was expected, but at the last minute I could not find anything about it in the paper. After stewing a while, I decided to go to the best source I could think of, looked up Mr. Whipple's name in the phone book (I was living in Cambridge), and dialed the number. Someone answered; I told whom I wanted; "Yes, this is Whipple," said the voice. I asked quickly, with apologies, whether the meteor shower was that night or the next. He was not the least put out by being used as an astronomical information booth, and spoke pleasantly for ten minutes about the meteor shower. Not much of a story, but it does tell something.

Since 1973 the cooperative science program of Harvard-Smithsonian has used the resources of both institutions, involving more than 120 scientists, and has also drawn liberally on the pool of talent in the Cambridge area. Observational instruments are backed up with libraries, computers, and laboratories. Many visiting scientists, post-doctoral fellows, and consultants join the Center for Astrophysics for a week or two, or more than a year. Work is currently under way in geoastronomy, high-energy astronomy, infrared and optical astronomy, radio astronomy, atomic and molecular physics, planetary sciences, solar and stellar physics, and theoretical astrophysics. Balloon-borne infrared telescopes are launched in Texas; analysis of X-ray data from the UHURU satellite goes forward; radio observations are carried out at McDonald's Millimeter Wave Observatory, at Haystack, and at the Agassiz Station; computer-simulations of stars and the solar system try to trace their evolution; and quantities of other studies are carried out on HAO-SAO telescopes or run on a competitive basis at Kitt Peak, Cerro Tololo, and elsewhere.

Owen Gingerich, of the Center for Astrophysics, writes: "You will be interested to know that we have hopes to restore the historic 15-inch telescope to its original condition and to maintain it as a museum of Harvard astronomy. Also, we are just in the process of erecting a 10-inch Fecker Refractor (owned by the Harvard Astronomy Department) on top of the undergraduate Science Center."

Visiting: Send a stamped, self-addressed envelope to "Open Nights," c/o Harvard-Smithsonian Observatories, 60 Garden Street, Cambridge, Mass. 02138. There are two series of "Open Nights" in late spring and early fall. Advance tickets are required for free admission.

In summer there are free observing sessions every Friday night using the refractors of the Cambridge observatory.

Accommodations: These are hard to find in the Boston area without spending a good deal of money. There is a chain called Susse [sic] Chalet Motor Lodges that offered singles at $10 in 1975; one of these is located at 800 Morrisey Blvd. at exit 19N off the Southeast Expressway, or 20S. Another is in Lexington off Exit 44N from Rt. 128, on Rts. 4 and 225. Further out Rt. 225 in Bedford there are several inexpensive places, including the Bedford Motel.

Nearby: The Peabody Museum of Harvard contains exhibits of great geological and astronomical interest. Many other observatories are listed in the Massachusetts section of Part II of this book.

United States Naval Observatory — Washington
Washington, D.C. 20390

Staff members of the Naval Observatory in Washington are in a peculiarly enviable position; you will understand why as soon as you drive or walk past the guardhouse at the gate and turn left up the hill towards the administrative offices. The beautifully maintained buildings scattered around the rolling lawns and stands of trees produce an effect somewhat like that of the Greenwich Observatory in England, and you feel that you are miles out in the country. Perhaps the situation will change somewhat with the Vice President now inhabiting the fine old frame admiral's mansion, but when I was there I couldn't see another tourist and scarcely another human being in any direction. Since it is a military establishment, only those on official business belong there, and those on "official business" usually include the specialized staff and a few visitors in the afternoons. The Washington traffic, which always was horrible and which during the construction of the subway has become unspeakably execrable, cannot be heard; the birds can. Despite the fact that the Observatory is readily accessible, there may be only three visitors there when, at the same time, there are a thousand in each of the downtown museums. If this happy situation continues, only the serious astronomical pilgrim will enjoy the most beautiful and peaceful surroundings in metropolitan Washington.

Political careers in this city wax and wane, but the stars remain eternally the same? No, actually the stars are far from being changeless, and the moon and planets shift

Aerial view of the Washington Naval Observatory. The Main Building is just above center; the white dome near the center houses the 26-inch refractor; the square buildings with white roofs just below center are roll-off housings for transit instruments. (U.S. Naval Observatory Photograph)

position hour by hour; it is a measure of man's own stability that he has been able to develop instruments the constancy of which is greater than that of the heavenly bodies. This took a long time; positional astronomy is the oldest branch of the science, which began with two giant steps widely spaced in time—Hipparchus and Tycho Brahe—and has developed with steady increments of precision since the mid-nineteenth century. Most observatories now no longer concern themselves directly with problems of astrometry, or star positions, and timekeeping; the Naval Observatory, along with Greenwich, England, and an observatory in the Soviet Union, is one of the few in the world that continually observes and redefines the positions of the sun, moon, planets, and stars. This is the only place in the United States where precise instruments gaze with a calculated naiveté at the skies and take the measure of celestial motions as if no one had ever done it before.

Since the position of a planet depends on the time that it is observed, the Naval Observatory is in many ways one big clock; it is the only observatory in the country that determines time, and the Naval Observatory's chronometers are those against which the National Bureau of Standards, whose WWV time-keeping signals can be picked up on any shortwave radio, checks its apparatus. The Observatory determines several different kinds of time, and for a more complete account of each you may study their well-written booklet. The technical name for our ordinary clock time is mean solar time; it is based on the rotation of the earth. In Washington and in another Time Service Substation in Florida, there are highly specialized telescopes called photographic zenith tubes. These point exactly vertically, and the position in the sky at which they are aimed depends solely on the latitude and the turning of the earth. To assure verticality, they use mirrors which are actually shallow pools of mercury; the mercury assumes a surface precisely perpendicular to the radius extending from the earth's center of gravity. When certain stars cross the meridian (an imaginary projection into the sky of the longitude at Washington, D.C.) they are photographed and the time is recorded by a clock. Examination of the photograph reveals if the clock was on time with respect to the earth's rotation. Because the earth is itself running "slow" (losing a little rotational velocity each year), it is sometimes necessary to adjust the time. The time signals are always within a second of agreeing perfectly with astronomical time, and the small difference between the two is published weekly; when the difference becomes too great, the clock drops exactly one second at the end of the month. The announcement of these "leap seconds" is sometimes made a good deal of in the press; you understand why there was such a furor when England dropped thirteen days from the calendar in the eighteenth century.

A visit to the time center of the Observatory lets you see the bank of clocks—about two dozen of them—which are constantly checked against one another and against the stars. Time-keeping technology continues to advance rapidly, and it seems likely that wrist watches with a chronometric accuracy that the Navy would have envied a few decades ago may soon be available for twenty-five dollars. Such watches and the Naval clocks depend on the very high frequency of molecular or atomic resonances and transitions. The watches use a quartz crystal; most of the clocks now used in the time center employ the oscillation produced by the cesium atom when an electron drops from one particular excited state to a state of lower energy. The frequency used has been defined and measured as equal to 9,192,631,770 cycles per second; there are precisely 86,400 seconds in a day; and the individual clocks are accurate to within 1/10,000,000 of a second per day. Taken all together, they keep time ten times better than that. In addition to the cesium clocks, which include portable models that can be carried to "deliver time" to various government agencies, there are hydrogen maser chronometers.

Determination of star positions is called astrometry, and some of its uses and techniques are described in the section on the Navy's Flagstaff station. At the Washington

The photographic zenith tube. It remains permanently in a fixed vertical position; note the mercury-pool horizontal mirror at the bottom. (Official Department of Defense Photograph by PFC Thomas J. Mackesy, USN)

The master clock complex in the time room. Each unit with handles is a cesium clock. (U.S. Naval Observatory Photograph)

Observatory the most precise instrument for purposes of planetary, lunar, and stellar positional astronomy is one of rather small aperture, which is not ordinarily shown to visitors: the 6-inch transit. To "transit" means, for a star, to cross that imaginary meridian directly overhead; hence this instrument is fixed with absolute rigidity in its north-south direction, immovable from east to west except by the earth's rotation. Massive pillars and huge foundations maintain the alignment of this small instrument, which must nevertheless be rechecked frequently in the course of observations by aiming it at an artificial star, a point source of light in a small building some distance to the north of the slotted structure from which the transit observes the narrow band of sky overhead. Precautions for assuring the accuracy of the observations become more elaborate with the increasing sophistication of auxiliary equipment. Military budgets are ample, and this small telescope—irreplaceable because from long experience its own quirks can be allowed for—is connected with a computer which constantly records and compensates (in the reduction of observations) for temperature, humidity, refractive index of the air at that time, height of the object and consequent refractive displacement, and even the individual observer on duty, whose native quirks of vision may lead him to indicate that the object has crossed the meridian a few milliseconds sooner or later than it actually has. Not really all that different from the pendulum clock and the telegraph key at nineteenth-century Greenwich; just more accurate.

When I saw the transit it was undergoing a rare period of "down time," when it is carefully cleaned and reversed in its bearings to equalize the microscopic amounts of wear.

The 6-inch transit circle. Note how movement is rigidly restricted to a north-south motion. The roll-aside roof slit is partly open; the wheeled couch supports the observer; a hand-held control paddle hangs on one pillar. (U.S. Naval Observatory Photograph)

Other instruments, more recently acquired, are a 24-inch Group 128 Cassegrain and a 15-inch astrographic camera that is useful for work on comets and asteroids.

What the visitor is privileged to see is one of the largest of the great nineteenth-century refractors. The 26-inch was built by Alvan Clark and Sons. They were commissioned in 1870 to construct the largest refractor in the world, to surpass the 25-inch just completed by Thomas Cooke in England. Unlike their practice in the construction of the earlier Dearborn refractor, whose original wooden tube may be seen at the Adler Planetarium in Chicago, the Clarks used riveted steel plates for the tube and placed it on a

The 26-inch refractor, built by Alvan Clark in the 1870s. (U.S. Naval Observatory Photograph)

mount whose weight-driven clock was wound by using pressure from the Washington water system. By means of the latest technology, the clock's errors were adjusted with an electromagnetic brake that acted on the pendulum. The Clarks also supplied equipment that enabled spectroscopic observations to begin, but the filar micrometers that were used for many years afterwards were the most important piece of auxiliary apparatus. These measure the relative positions and separations of binaries, "double stars" connected to one another by gravity and revolving around one another. Determination of an orbit may take many observations over many years, but once it is achieved, other information, such as the weights and even the distance of the pairs of stars can be estimated. Widely separated binaries can be photographed, but closer pairs require actual visual observation, one of the very few instances in which the eye outperforms the camera. The definition of this long-focus refractor makes it ideal for such work.

The refractor has undergone two extensive remodelings since its installation. In 1893 it was moved to its present site from the banks of the Potomac where swampy areas produced occasional miasmas. A new mounting by Warner and Swasey replaced the old one, which was too weak. That company also built the dome, which rests atop the white marble building named for Asaph Hall, the second director of the Observatory and the discoverer (in 1877) of the two satellites of Mars. Early in the 1960s many of the mechanisms in the mount were modernized, motors to supply fast motions were added, and the mechanical clock was replaced by an electric drive. An important change that led to the resumption of binary observation was the addition of digital recording devices to the filar micrometer, making readout much quicker and more accurate. Although seeing conditions are very bad, it is still possible to carry out more than a thousand observations in a year.

The dome in which the Clark instrument is housed is itself an object of beauty. All surfaces of the instrument and the dome interior are kept painted white, immaculate as a dress uniform. The hardwood of the elevator floor (a real luxury for a refractor observer) gleams beneath its varnish like a ballroom. Anyone who wishes to impress dubious friends or relatives with the worldly dignity of astronomy should bring them here.

Visiting: A one-hour tour of the Observatory begins at 2 p.m. Monday through Friday, except on holidays. Remember that this is a time center for the world, and that the tour will be under way ten seconds after the hour. Evening tours three nights per month, places in which must be reserved in advance by calling (202) 254-4569 in Washington. The gate to the grounds is at 34th Street and Massachusetts Avenue. Do not be deterred, but turn in at the guardhouse and explain what you are there for.

Accommodations: One hardly need say that Washington is an expensive place to stay, though travel guides may help let you know what you are getting into. The only definite suggestion here is the Rock Creek Hotel. At 1925 Belmont Road and 20th Street, one block east of Connecticut Avenue, it is less than a mile from the Naval Observatory. In late 1974 it had singles for $12 and doubles for $14. It operates strictly on a cash-payment-night-by-night plan and will not take reservations; but it is clean and comfortable, with a lot of free parking space. Breakfast in the attached coffee room is cheap and convenient but is not recommended. To get a room, show up in person early in the day.

Nearby: The Smithsonian's Museum of History and Technology must be visited by anyone interested in astronomy. They have a huge hoard of historical instruments, to which they add constantly, and many of these are placed on display in intelligent and imaginative fashion. Telescopes and reproductions of telescopes of all eras, actual mirrors and lenses by the great names in the optical profession, a full-size reconstruction of Henry Fitz's telescope-making shop, a Foucault pendulum, and much more make this the world's finest exhibit of astronomical optics. Allegheny, Green Bank, and Leander McCormick Observatories are all within a half-day's drive of Washington.

Leander McCormick Observatory
Charlottesville, Virginia 22903

In 1872 Leander J. McCormick, son of the inventor of the reaper, put in an order with Alvan Clark and Sons for a refractor of aperture equal to that just installed in the Washington Naval Observatory. McCormick had been born in Virginia, and hoped to endow its state university with one of the great telescopes of the world—and had things proceeded on schedule the Observatory would have tied with the Navy for the largest refractor in existence. Business cycles in the nineteenth century were more extreme than now, though, and like other astronomy-loving tycoons, McCormick found himself short of cash for a while. As Deborah Jean Warner points out, "This delay gave them a chance to learn from some of the mistakes of the Washington instrument. The inner surfaces of the Virginia 26-inch objective, for instance, were given slightly different radii, so as to avoid annoying 'object-glass ghost.' The driving clock was connected with a Seth Thomas located in the computing room of the observatory." In 1883–84 the Observatory was finally completed and the telescope took its place atop its brick pedestal.

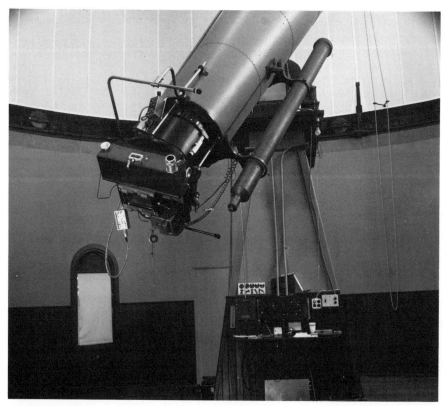

Tailpiece and guiding controls on the University of Virginia's 26-inch telescope. (Leander McCormick Observatory Photograph)

Since that time, the 26-inch has contributed thousands of the parallax determinations essential for establishing the first mileposts out into the galaxy, and many more plates showing proper motions. In 1973 the mounting and pedestal were deteriorating somewhat; one hopes that funds earmarked for renovation of this great telescope will turn up.

Charlottesville, with a population of 40,000, had by the early 1960s become too great a source of light pollution, and the Department of Astronomy began to look for a better site for any new telescopes that might be constructed. Although the Blue Ridge Mountains lie a short distance to the west, tests showed that when weather systems moved over them they produced very turbulent air patterns. The site that was chosen was Fan Mountain, an isolated peak 16 miles south of Charlottesville which rises 1,819 feet above sea level. Its slopes are covered with trees, which help stabilize local thermal conditions. The Astronomy Department claims that, "In terms of transparency, seeing, and especially sky darkness, Fan Mountain is one of the best observing sites available on the east coast." It is convenient to Charlottesville; the chartered bus that carried the members of the American Association of Variable Astronomers to Fan Mountain in 1973 made it from the city limits to the base of the mountain in a breakneck 15 minutes. The gravel road up is steep and curvy, but the top flattens out and provides plenty of space for any number of installations.

The largest instrument currently at Fan Mountain is a 40-inch Cassegrain Schmidt astrometric reflector. This takes plates 10 by 10 inches, each covering a field of 1 degree. In order to make the most use possible of the plate collection accumulated with the 26-inch refractor, the two telescopes have been photographing identical star fields simultaneously. This will help to detect any anomalies in the plate scale of the 40-inch, so

Aerial view of Fan Mountain Station. (Leander McCormick Observatory Photograph)

The 40-inch astrometric Cassegrain-Schmidt at Fan Mountain. (University of Virginia, Dept. of Graphics Photograph)

that such irregularities can be allowed for in reducing the data for proper motions and parallaxes. As with the Naval Observatory's astrometric reflector at Flagstaff, astrometric astronomers elsewhere worry whether an astrometric instrument with movable optical parts can be completely reliable. This experiment should tell a lot, with the 26-inch providing a standard for the 40-inch.

The first large telescope installed at Fan Mountain was the 32-inch Tinsley Cassegrain. This can be used at f/16 and f/32 for photometry and spectrography. Auxiliary equipment permitting such studies consists of: single- and dual-channel photoelectric photometers with DC and pulse-counting data systems and filters for broad- and narrow-band photometry, a photoelectric spectral line scanner, and a Cassegrain spectrograph equipped with image intensifiers. In the past, the 32-inch instrument has been used with wide-band and narrow-band interference filters for photometry of magnetic stars and UBV photometry of eclipsing binaries. Current interest seems to be in instrument development, with experiments in image intensification, one-micron spectroscopy, and photoelectric astrometry. The 32-inch telescope is aimed and guided by a movable diagonal mirror and a movable eyepiece in a section behind the main mirror called the guide box. There is also a 5-inch Tinsley-made Maksutov finder, so compact and attractive that numerous amateurs have coveted it for their main instrument.

A separate small dome houses the 10-inch astrograph that operated in Charlottesville for twenty years (1935–55) making an objective-prism survey of the stars in the northern sky down to magnitude 12. This large-scale camera takes plates of 10 by 10 inches, which cover a square area of 7 degrees on edge. Now back in use, the astrograph has contributed to the almost 5,000 low-dispersion plates in the collection of the astronomical research library.

Supporting equipment and laboratories for the Observatory instruments are extensive. With such a large collection (over 100,000) of astrometric plates and photographs of stellar spectra, one is not surprised to find that there are numerous measuring tools: a digitized, two-screw Grant Stellar Comparator, a Mann Stellar Comparator, and several Gaertner comparators. There are an optics and photocathode laboratory, a large microdensitometer, and shops for machine and electronics work.

The Department of Astronomy at the University of Virginia has close ties with the National Radio Astronomy Observatory, which has offices and computers, and carries out theoretical studies, right on the campus. Graduate students in the department can arrange to do work for their dissertations under the guidance of the National Radio Astronomy Observatory staff. For descriptions of NRAO installations, see the entry on Green Bank in Part I of this book, and the account of the Very Large Array under New Mexico, in Part II.

Much of the work at Virginia is theoretical, focusing on stellar structure and evolution, and on cosmological problems of the largest scale. The program ranges from the most traditional observational disciplines to the most advanced speculation. I was privileged to hear a lecture by William Saslaw (at that time the youngest full professor at Virginia) which detailed an imaginary journey to the center of an exploding galaxy, and equally privileged to hear the congratulatory joshing of Charles Tolbert: "Well, Bill, that got right scary at times!"

Visiting: McCormick Observatory, on Mt. Jefferson at the edge of the university campus in Charlottesville, is open every first and third Friday night each month from about 7:30 to 11 p.m. There is a lecture, a movie, and observing through the 26-inch refractor if the sky is clear. Views of planets through this instrument, when the air is steady, cannot be surpassed. About 5,000 people a year take advantage of these evenings.

Three times a year, as announced one month in advance, the Fan Mountain Station is open for public visits. Roughly, these dates occur on the third weekend in October, April, and July.

Both the gravel road up Mt. Jefferson and the last stretch of steep gravel road up Fan Mountain (off US 29, about 15 miles from the southern city limits) are a little hard to find, especially in the dark. Campus maps may be obtained at a University of Virginia information office; the administrative office number of the Observatory is (703) 924-7494, which you might call to ask about landmarks if you are going to an open house at Fan Mountain.

Accommodations: The Siesta Motor Court, 7 miles west of Charlottesville on US 250, had singles for $10 in 1975. This is also on the way from or to Green Bank. The Mount Vernon Motel (at US 250 bypass and US 29 N) is a large establishment with slightly more expensive rooms, as is the Downtowner near the University of Virginia campus. Halfway to Washington, at the Boxwood House Motel in Culpepper, there were singles for $9 in 1975 and other even less expensive places might be found along US 29.

Nearby: It is not hard to find an excuse for mentioning Monticello, since Jefferson interested himself in all the arts and sciences, owned a small telescope, and devised a calendar clock in the hallway of his mansion.

Since Charlottesville is located at the halfway point between Washington, D.C., and Green Bank, West Virginia, visits to all of these places might be included in a single trip. See descriptions of the Naval Observatory and Green Bank.

Green Bank:
The National Radio Astronomy Observatory
Green Bank, West Virginia 24944

As you turn in from the highway to the Observatory at Green Bank, you see three ungainly contraptions on either side of the road. One of these in particular, with its Ford

Model T wheels and its flimsy tubing, looks as if it might have been set out for the next trash collection—a piece of defunct kinetic sculpture. Actually this is a replica of the antenna that Karl Jansky set up in 1932 at the Bell Laboratories in Holmdel, New Jersey; when Jansky discovered that he could regularly pick up random radio noise from the Milky Way, he had invented radio astronomy. He had been trying to account for something that interfered with telephone communications, and had discovered that it was the galaxy itself that caused the problem.

Replica of the Jansky antenna; note the Model T wheels. (NRAO Photograph)

Since the beginning, the problems of radio astronomy have been in large part problems of astrometry, the location in the sky of radio sources. Jansky could only tell that certain intensifications of radio noise coincided with the passage of the Milky Way, which is, of course, by far the largest extended source in the sky. His "telescope" had an angular resolution of tens of degrees at best. Two other displays near the gates illustrate stages in achieving more precise location and identification of sources: the Original Reber Radio Telescope, built in 1937, and the Ewen Horn. Grote Reber built his 31-foot paraboloid in Wheaton, Illinois; the National Radio Astronomy Observatory acquired it in 1958 as a historical exhibit, but the alt-azimuth instrument was also actually used for research.

The working equipment of NRAO is operated by a group called the Associated Universities, Inc., under contract with the National Science Foundation. These facilities are open to any qualified scientist or graduate student, who may submit a proposal for work to be carried out with one of the instruments. Of available observing time on the telescopes, 60 percent is assigned to visiting observers. In 1973, 217 visitors from fifty-five different institutions used the equipment; fifty-nine of these were students. The staff of the

The original Reber radio telescope, installed as an exhibit but still usable. Reber built this in his back yard as an amateur project. (NRAO Photograph)

Observatory, whose main office is located on the campus of the University of Virginia in Charlottesville, cooperates with the Department of Astronomy there and with the Leander McCormick Observatory (q.v.), in the training of graduate students.

One of the Green Bank telescopes compares favorably with any in the world for sheer size; this is the 300-foot meridian transit dish. At the time of its construction, it was decided to make the instrument steerable only in a north-south direction; the available funds were committed to gain collecting area rather than complete steerability. As a consequence, the telescope, like the transit refractor at the Naval Observatory in Washington, must be set at a particular declination and then wait for the passage of the object that is being observed, in effect using the earth's diurnal motion as a steering mechanism. In the spring of 1973 the meeting of the American Association of Variable Star Observers adjourned from Charlottesville to Green Bank; when we arrived at the 300-foot instrument, there were a few minutes of leeway before the arrival of the next radio source on the meridian, so the people running the telescope kindly tipped it as far as it would go to the south so that we might get a good look into it. As the gears grumbled and the mammoth link chain tilted the big dish over, I felt especially favored; seldom has a piece of apparatus as big as a football field in all directions tipped respectfully in my direction. A few minutes later the antenna was back up awaiting the arrival of a radio source, and we were inside watching; the world wheeled on, the pens began to twitch on the chart paper, and in a minute the observation was complete. The main uses of this instrument are for the mapping of hydrogen in the galaxy (the method that first gave a direct picture of the spiral structure of the Milky Way), for pulsar research, and for studies of the radio continuum between 2- and 40-centimeter wavelengths.

The 300-foot transit telescope at the National Radio Astronomy Observatory in Green Bank. The control room is to the left. (NRAO Photograph)

Inside the control room of the 300-foot radio telescope. Banks of computers, radio receivers, and recording equipment make up the "eye" at the receiving end of such telescopes. (NRAO Photograph)

The second-largest telescope is the fully steerable, equatorially mounted 140-foot dish. Stabilizing the surface of such a telescope against mechanical deformations in different attitudes and against wind pressures is a considerable problem. But the flexibility of being able to aim at one spot in the sky, and track it, make the engineering worthwhile. The 140-foot is continually improved by the addition of new equipment; in 1973 it was converted to Cassegrain operation with a multifrequency "front end" operating in four bands, from 1350 megahertz to 24 gigahertz. When Comet Kohoutek appeared, the 140-foot tracked it for radio emissions, and the telescope has been used for many other purposes, including part of a Very Long Baseline Interferometric Array. For these last experiments, it has been used with various combinations of other radio telescopes such as an 85-foot dish in Sweden; the Haystack 120-foot in Massachusetts (q.v.); Owens Valley, California, telescopes (q.v.); the 1,000-foot Arecibo, Puerto Rico, telescope (q.v.); and others in the United States and Europe. One method is to make simultaneous observations of the same object and record them on tape, marking time with a very stable independent time source such as a hydrogen maser or a cesium clock. The tapes then are played against one another to produce interferometric "fringes" and in effect turn much of the earth's surface into a giant radio observatory. Such enormous "aperture" makes possible more accurate determination of diameters and structure of many kinds of radio sources, and also studies of the surface of Venus with much greater resolution than previously achieved. The speed with which the technology of radio astronomy is developing, and the daily discovery of new sources (or refinements of observations of older ones) can, in their implications for astronomy, be compared to Galileo's first hours with his first telescope.

The interferometric array of three 85-foot dishes; the two in the foreground are movable for aperture synthesis. Note the farm buildings left in place. (NRAO Photograph)

33

Strung out in a line across the valley floor are three other large parabolic reflectors, each 85 feet across. Two of these equatorially mounted dishes are movable along tracks for the purpose of aperture synthesis; by changing the separation of any two elements of this interferometric array it is possible to make them behave like different segments of a larger antenna and to build up a radio picture of a source. These can also operate in conjunction with a 45-foot antenna that is set up 27 miles away to achieve even larger aperture, and the whole assembly participates in the same sort of Very Long Baseline studies as the 140-foot antenna. The major work of the three 85-foot dishes has been in establishing source positions and in mapping small scale structures. In some ways it has served as a prototype system for the Very Large Array, also being built by the NRAO, on the Plains of San Augustin in New Mexico (see Very Large Array under New Mexico in Part II).

The Observatory has a 36-foot antenna at Kitt Peak; the dry air there permits observations in the millimeter-wave band, a region of the electromagnetic spectrum that tends to be absorbed by water vapor.

Why locate a great radio observatory in the eastern United States, and why in the mountains of West Virginia? As soon as you drive into the valley you sense part of the explanation: this beautiful plain—with silos still standing amid the radio antennae—is ringed by mountains that help cut off radio transmissions from human sources. The FAA has cooperated to route flight paths away from the area and the FCC has limited the types of radio equipment that can be put in use in the area. Artificial satellites still pass over regularly, and their transmissions are particularly noisome to the radio astronomers; but that would occur anywhere on earth. Gray skies do not hinder most radio-frequency work. Wind velocities are lower in the region than almost anywhere else in the country. Low population density reduces the likelihood of taxicab radios or CB communications. Yet an hour or two away is Charlottesville, Virginia, and Washington, D.C., is only a hundred miles further. Readily accessible by automobile, there is little through traffic in the valley, though, and radio waves produced by car ignitions are minimal; within the Observatory grounds the vehicles are diesel-powered, eliminating spark plugs.

Visiting: There is an excellent visitors' program here, and no itinerant astronomer should pass it up. From mid-June through Labor Day there are daily tours, given on the hour from 9 a.m. to 4 p.m. In September and October there are weekend tours on the same schedule. Without charge you will see a short film on radio astronomy and then be taken on a tape-narrated bus tour with stops at the principal operating radio telescopes. Exhibits along the way include an operating 2-foot telescope that tracks and monitors the sun, plus informational panels at each of the stops. The historical displays already mentioned are near the gates. There is a tour center where one may purchase popular books on radio and optical astronomy. To reach Green Bank, take SR 39 from I-81 to Minnehaha Springs, across the mountains, and then turn north on SR 28/92; or head west on US 250 (a continuation of I-64 from Charlottesville) until you reach SR 28 and turn south. The roads are full of curves and the trip takes longer than it looks on a map.

Accommodations: Not much in the way of motels in the immediate area, though there are some establishments along the roads mentioned; Warm Springs to the south is an old (and somewhat strange) resort town with a lot of motels and rooming houses. See Leander McCormick Observatory for specific recommendations near Charlottesville. There are a lot of campgrounds in the area; within a few miles to the west of the Observatory are the Whittaker and Leatherbark Campgrounds at Cass, West Virginia; there are other beautiful campsites both to the north and the south. High rainfall makes this not a particularly good region for optical telescopes.

Nearby: A camping trip or a vacation to this area could with little effort include visits to Leander McCormick Observatory, to the Naval Observatory in Washington, and to the astronomical section of the Museum of History and Technology. Less than 200 miles directly north of Green Bank (albeit over impossibly winding roads) is Pittsburgh, with the great Allegheny Observatory and the Buhl Planetarium. Except for Leander McCormick, all these places are open most days during the summer.

Allegheny Observatory
1935211
University of Pittsburgh, Pittsburgh, Pennsylvania 15214

A person asked to select the unlikeliest site in the United States for a major observatory might well include near the top of his list Cleveland and Pittsburgh, yet Cleveland harbors at least three fine observatories, and on a hill above Pittsburgh the third largest refractor in the country collects fundamental data about the stars. Only a few years ago, though, the Allegheny Observatory almost did get closed; fortunately, the opinions of professional astronomers (and also of the concerned citizens of Pittsburgh) were listened to, and the state government appropriated $25,000, so that this attempt did not succeed. As in the case of Yerkes, Leander McCormick, and the Naval Observatory in Washington, the very continuity of observation is a resource that cannot be duplicated by the newest equipment in the best locations. One cannot travel back in time sixty years at Kitt Peak, photograph the heavens as they were at that time, and compare the results with a photograph taken today with precisely the same optics; one can do that at Allegheny.

From the beginning, the history here has been a history of personalities. Many children in Pittsburgh, for example, are brought up to revere the memory of John A. Brashear, maker of the Observatory's two largest telescopes, as a kind of patron saint. And indeed anyone who suspects that the homage is overdone ought to read Brashear's autobiography, which describes his long and dangerous days as a Pittsburgh millwright. His career as a lens-maker began with his efforts to make a 5-inch refractor at home; despite the fact that he was a genius with mechanical things, it was a lens-breaking and heart-breaking experience. All Brashear's self-instruction in optics, and the tedious hours of grinding and figuring, took place after what was often a twelve-hour day at the mill. The picture of him and his wife sitting up late, night after night, in the little shop behind the house, ought to dry the eyes of any graduate assistant tempted to self-pity in these days. It is a happy coincidence that the second director of the Observatory was Samuel Pierpont Langley, himself a persistent and inventive man; when Brashear made his timid overtures, Langley was kind enough to praise his first efforts in optics, and to allow him to look through the Observatory telescope.

As his techniques improved, Brashear began to get orders to make lenses for other people, and for a while he tried to do this and also to hang on to his job at the mill. He had reached a state of nervous collapse and profound indecision when William Thaw, a wealthy Pittsburgh citizen, offered to set him up properly in the business of producing fine optics. Thaw provided a building and equipment which Brashear was to use without charge until he could make his own way. The industrialist must have been a keen judge of talent; he had already helped Langley with the Observatory, and Langley went on to the highest scientific post in the country, secretary of the Smithsonian Institution. This time Thaw helped the man who was going to make some of the largest lenses and mirrors in the world, including the 30-inch at Allegheny called the Thaw refractor.

Another great name associated with the earlier days of the Allegheny is James E. Keeler. Working with the original 13-inch refractor (still used almost nightly for visitor viewing) Keeler demonstrated that the inner and outer parts of Saturn's rings revolved differentially, and must therefore be made of small particles; he did this by observing different Doppler shifts for the different areas of the rings. After he went to be director of Lick Observatory in 1898, he continued his spectrographic observations, adding

The Thaw refractor, 30-inch aperture, of the Allegheny Observatory, made by John Brashear in 1912. The opening in the pier reveals a centrifugal governor for the clock drive; several galleries provide access to a rising floor at different levels. (Allegheny Observatory Photograph)

enormously to astronomical knowledge in the brief two or three years before his death. It was, incidentally, at that time that he met Alvan Graham Clark and found him to be such a grouch; perhaps Keeler was too used to the lovable John Brashear.

The Allegheny Observatory took its present form in 1912; previous to that it had been located in what is now the near Northside of Pittsburgh, and consisted of a 13-inch refractor made by Henry Fitz, installed in an onion-shaped dome in 1861. This telescope rode into existence on the crest of a "comet boom," since the bright comet of 1859 had

36

inspired the formation of an Allegheny Telescope Association. The excitement subsided, and in 1863 the telescope came into the hands of Professor Philetus Dean, its first director. Dean became notorious for defending the instrument against all prospective viewers, apparently judging it much too valuable to be put into use. One wonders what he would think if he knew that several thousand people a year now look through that same lens. At any rate, he made such a mystery of the telescope that in 1872 the lens was kidnapped and held for ransom. By that time Langley was director, and after standing firm he got the lens back; he met the lens-napper in some woods and was told by him, "You are a gentleman and I am a gentleman," as the thief handed it over. The 13-inch lens was slightly damaged; this had the happy consequence of its being worked over by the Clarks. In the phrase, quoted by Deborah Jean Warner, of a French nineteenth-century work on observational astronomy, it was now, "l'un des meilleurs objectifs de cette dimension que l'on connaisse."

As Pittsburgh grew, a new Observatory site became increasingly necessary; over a period of more than fifteen years, with fits and starts that often had to do with general economic convulsions, it took shape. First, Brashear, who had become chairman of the Observatory Committee, found a hilly site some miles north of the city and above much of the smoke. Construction began in 1900, and in 1906 a reflector of 31-inch aperture, made by Brashear and named for Keeler, was installed. This can be used as a prime-focus instrument, a Newtonian, or a Cassegrain. Its most remarkable feature, though, is a coelostat-fed high-dispersion spectrograph, one of the earliest such systems ever constructed. The Porter Grating Spectrograph, for solar studies, is actually located several floors below the telescope in the Solar Tower. With a modern aluminum coating, Brashear's mirror still serves well for spectroscopic studies of binaries too close to separate visually; for this purpose it uses the Mellon Doppler spectrograph attached at the Cassegrain focus.

When Brashear was designing the elements for the 30-inch refracting lens, completed in 1912, it had become obvious that the most important future work would be done with photographic plates. The curves were ground, therefore, to achieve the sharpest focus towards the blue end of the spectrum where silver-based emulsions are most sensitive. This design, coupled with the relative steadiness of star images in the area, help make up for the pollution by light and smoke. Add to this the practice of making astrometric plates with the telescope always to the west of its pier, so as to duplicate all the slight stresses on the objective, and you get some of the sharpest photographs of star fields available. According to the acting director, Joost H. Kiewiet de Jong, star images are sufficiently well defined to permit measurements of a twenty-thousandth of an inch. At this writing the whole instrument is being renovated, with new remote slewing, setting, and circle readout systems being installed. The lens is being checked over by the Fecker Systems plant, a linear descendant of Brashear's own company.

Various kinds of auxiliary measuring equipment—including a measuring machine made available at the Washington Naval Observatory—are continuing the examination of the plates and spectra that accumulate.

Discussions of the sort of fundamental work that the Observatory pursues in astrometry may be found in the descriptions of the Naval Observatory Flagstaff Station and of Yerkes. And in common with another large Brashear telescope, at the Sproul Observatory, the Thaw refractor searches for stars whose oscillations betray the presence of an invisible companion.

Visiting: The Allegheny Observatory is perhaps unique in having an endowed visitor's program; this was made possible by Henry C. Frick and his daughter Helen. It has been staffed by graduate students since 1972 and operates six nights a week (except Sunday) from April 1 until early November. The program includes lectures, slides, movies, and visual observation through the 13-inch refractor. So experienced is the Observatory in serving the public that in 1966 two of its astronomers, Mullaney and McCall, produced a guidebook to the heavens, *The Finest Deep-Sky Objects* (Sky Publishing Corporation). Once a year all three telescopes are made available to the public at an open house. Amateur astronomers are free to set up their own instruments on the Observatory grounds. Daytime visits can also be arranged; in 1974 there were over a hundred such groups received, totalling 2,589 visitors.

Accommodations: Although there are many motels in and around Pittsburgh, representing all the big chains, prices start at about $15. To the north on US 19, though, especially as you approach and pass I-94, there are some motels that are a few dollars cheaper; since the Observatory is on the north side of Pittsburgh, access is not all that bad from these places.

Nearby: The Buhl Planetarium and the Carnegie Institute's Museum of Natural History are two fine institutions for popular education in astronomy and the other sciences that are located in Pittsburgh. Pennsylvania is actually heavily infested with astronomical institutions, many of which are listed in Part II under the state's heading.

Yerkes Observatory
Williams Bay, Wisconsin 53191

The Yerkes Observatory is probably too attractive to visitors for its own good. Cool summers and the beauty and conviviality of Lake Geneva make the whole region a retreat

The main building at Yerkes. The 40-inch refractor is in the large dome. (Yerkes Observatory Photograph)

from Chicago, and a national convention center. Except for Greenwich Observatory, and possibly the Naval Observatory in Washington, no telescope is housed on lovelier grounds. I saw it in the snow, but those expanses must stretch beautifully in summer— acres of grass and evergreens, with shrubbery around the main building. Romanesque-revival architecture was never so happily employed; it suits observatory domes perfectly. The largest dome even resembles the baptistry at Pisa. Not since Tycho Brahe got his hands into the royal treasury of Denmark has anyone built so ideal a Stjernberg. When you add the celebrity of the great refractor to these other beauties, you wonder if the Observatory's claim of 15,000 visitors per year is not understated.

The birth of the Yerkes Observatory is another of those stories of chance, strong personalities, scientific curiosity, public relations, and money. George Ellery Hale figures in several such episodes; this was his first great triumph. In a casual conversation with Alvan Graham Clark at a meeting of the AAAS in Rochester in 1892, Hale heard about two 40-inch discs that had been cast by the firm of Mantois of Paris. With the collapse of a land boom, the University of Southern California, which had originally ordered the discs, found that land donated to pay for a planned telescope was no longer valuable enough. Hale, who already had a 12-inch refractor in his own observatory, went to President Harper of the University of Chicago with his story about the huge discs, which the Clark telescope firm was prepared to figure. Harper told Hale to show his plans to Charles T. Yerkes who had made a great deal of money building up the Chicago trans-portation system. At first Yerkes would only commit himself to paying for the finishing of the lenses (the Clarks themselves had a $20,000 investment in the optical blanks). Over a period of years, though, Hale managed to get Yerkes to pay for the whole thing, includ-ing the rising floor.

Two incidents delayed the dedication of the building in 1897. One trustee objected to a particular carving on the columns that flank the main entrance: there was a man being attacked by a swarm of bees, a device that occurred eight times. The trustee thought it undignified, so someone had to take a hammer and chisel and knock off the bees. More seriously, the rising floor, which weighs almost forty tons, snapped some of its cables and collapsed the night before the trustees and other officials were supposed to walk out onto it for the ceremony.

As a booklet published by the Observatory says, "The 40-inch objective is the largest lens ever constructed and successfully used for astronomical observations." Somewhere in the Paris Observatory, in storage, there is a 50-inch lens, but it was never used for anything but a horizontal telescope fed by a coelostat, set up as a crowd-pleaser at an international exposition in 1900. It is usually said that the Yerkes lens represents the largest possible in a refracting instrument, because anything larger would be seriously deformed under its own weight. But others have argued that such deformations occur in a manner that automatically compensates for itself, and Deborah Jean Warner wrote: "Had Alvan Graham lived longer—he died in June 1897, shortly after delivering the objective to the observatory at Williams Bay, Wisconsin—lenses of even greater aperture might have been made. The often mentioned problems of adequate support and absorp-tion of light through the glass seemed trivial to him."

The vital statistics of the great refractor are spelled out in G. Edward Pendray's *Men, Mirrors, and Stars* and in a pamphlet available from the Observatory which reproduces some of that book verbatim. Briefly, the front and back lens elements are separated by 8 inches, and are thinner than one might suppose possible: somewhat more than 2 inches at the thickest; the tube is over 60 feet long; the mounting rests on a four-section cast-iron pier which is in turn on a foundation of brick and concrete (which you can see by peering under the edge of the floor). The rising floor has a range of 23 feet; it is

The largest refracting telescope in the world. (Yerkes Observatory Photograph)

supported by huge counterweights between shafts up and down the walls. The 40-inch is the greatest astrometric instrument in the world. According to an astronomer-guide that I heard, the lens cell has never been disassembled, and therefore the relation of the optical surfaces now is identical with that when the first plates of star fields were made. This means that measurements with stellar comparators give proper motions with an accuracy obtainable in no other way.

In its early years, the telescope obtained photographs of dim objects with detail never seen before. One of these, of the Orion Nebula, actually passed through our hands during our visit; since the fifty sets of fingers that handled the thin glass plate were all either numb or in gloves—the unheated dome being somewhat below freezing at the moment—this was almost as much a trial of nerves as a privilege!

Another most important early piece of work was the determination of many stellar parallaxes.

Yerkes is designed to operate, as nearly as possible, as a self-contained astronomical institution. To supplement the work of the great refractor there are two other large instruments, located in the smaller domes of the main building. Proper motions need radial velocities in order to determine the true motion of a star with respect to the earth; the spectroscopic equipment used with a 41-inch Cassegrain-coudé provides this information. This telescope can also be used for image-intensifier-assisted spectroscopy, photoelectric photometry, and various other purposes. In the small northeast dome, there is a 24-inch Cassegrain whose tube rotates around its optical axis in order to help measure polarized light; this gives more precise results than simply rotating a filter. A very sensitive photometer attached to this reflector uses a fast counter to register single photons as they are collected by the telescope.

The extensive supporting facilities include a newly built climate-controlled plate vault with space for measuring engines; also, there are machine shops, lecture rooms, optical and electronic shops, offices, and computers.

Astronomers associated with Yerkes and the University of Chicago carry out projects in all branches of astronomy. Yerkes still has a close connection with the McDonald Observatory (q.v.) whose 82- and 107-inch reflectors and superior seeing conditions permit advanced astrophysical research not possible in Wisconsin. The astronomers also secure time on the optical and radio telescopes at Kitt Peak, Green Bank, and in the southern hemisphere observatories. They work with NASA to plan observations with Lear-jet-borne telescopes, and orbiting space telescopes. One only need mention that Chandrasekhar is on the faculty to show that theoretical work is also of great importance here. A dozen or so graduate students do course-work or dissertation projects at the Observatory.

Visiting: In winter, between October 1 and May 31, the Observatory is open only on Saturday from 10 a.m. until noon. Lectures are at one-half hour intervals, or as soon as enough visitors accumulate to make it worthwhile. In summer, June 1 to September 30, it is again open only on Saturday, 1:30 to 3 p.m., with lectures at half-hour intervals. Visitors are conducted into the dome of the great refractor and sit around the edges for the lecture. The dome is unheated in winter, to prevent convection currents at night. The hallway of the main building contains some dingy and outdated transparencies and some badly labeled exhibits of meteorites, old instruments, etc. The drive from Chicago takes two hours and the best way to go is up I-94 to SR 50. US 12 is shorter, but it passes through numerous suburbs, with frequent stops for traffic lights.

Accommodations and Nearby: Williams Bay and Lake Geneva are very popular summer resorts. Unless you are the sort who can bargain successfully for a room in the off-season, it would probably be better not to try to stay here. There is a campground, if you can find a space in it, at Big Foot Beach, off SR 120 south of Lake Geneva. Chicago is not cheap either, but the Heart O'Chicago Motel had singles starting at $12.50 in 1975; doubles a few dollars more. This motel is at 5990 N. Ridge Avenue, near the lake shore on the north side of town, where US 14 angles from Ridge onto Patterson. This is actually a good location for the astronomical traveler: 1. Evanston, with the Dearborn Observatory and the Lindheimer Research Center, is just to the north; 2. Access is not bad to the Lake Shore Drive to the south, where the Adler Planetarium (whose astronomical museum is one of the best in the world), the Field Museum, and the Museum of Science and Industry can be easily reached; 3. A continuation of US 14 takes you up the North-west Highway, where at 5700 you find the American Science Center; of all surplus optics and science places that I have searched out—in London, Paris, Washington, New Jersey, Pasadena, and Los Angeles—this is the best, with actual surprise bargains in giant eye-pieces, etc. Another overnight location 45 miles from Yerkes might be Madison, Wisconsin. Two Roadstar Motels offered singles at $8.88 in 1975; one is located five miles northeast of the center of town on SR 151; the other just east of I-90 on US 12/18. Madison is the location of the Washburn Observatory (q.v.).

McDonald Observatory
at Mount Locke, Fort Davis, Texas
(University of Texas at Austin 78712)

A traveler approaching the McDonald Observatory from the east will find his arrival among the Davis Mountains a welcome break from the hundreds of miles of plains that he has just crossed. Those mountains are there because of great lava flows, which are in places thousands of feet thick. The flows make the cliffs—tall columnar joints weathered brown—that flank the road and provide a theatrical backdrop for the restored garrison post at Fort Davis.

McDonald Observatory occupies the summit of Mount Locke, one of the multilayered volcanic remnants, 6,800 feet above sea level. Light pollution at this site is minimal; the

Summit of Mount Locke. The 107-inch telescope is housed in the dome in the foreground, and the 82-inch Otto Struve reflector in the larger dome on the left. (McDonald Observatory Photograph)

area has few people living in it, and the nearest large city, El Paso, is 170 miles to the west. Here in southwest Texas there are twelve more degrees of southern sky to be seen than in Wisconsin; for this reason it is not surprising that the University of Chicago astronomers at Yerkes were anxious to cooperate in the establishment of a large telescope. This occurred in 1932, when W. J. McDonald left close to a million dollars to the University of Texas for the building of a large telescope. A cooperative program was worked out whereby the Yerkes director would also direct McDonald, since at that time the University of Texas had not built up the strong Department of Astronomy that it now has. Although Yerkes still has close ties with McDonald, the facilities on Mount Locke are now operated by the University of Texas.

William Johnson McDonald was another example of the type of man that has served the same purpose for American astronomy as the Medicis did for renaissance art in Italy. As the bronze panel at the Observatory by the Texas Historical Survey Committee puts it: "A Paris (Texas) banker interested in the stars. A well-educated man, McDonald lived frugally. As a hobby, he read science books and viewed planets through a small telescope. His will granted to the University of Texas $800,000 'To build an observatory and promote the study of astronomy.'"

When the 82-inch telescope was completed, it was second in size only to Mt. Wilson's Hooker Telescope. It is a versatile instrument, with prime focus, Cassegrain, and coudé arrangements available to the observer, and has been used for most kinds of optical astronomical work: photography, photometry, spectroscopy, polarimetry, and so on. For a time it was the most powerful instrument for measuring the red-shift of galaxies at the limits of the observed universe; several new planetary satellites were found with it.

A few years ago this telescope was the object of a lunatic assault, when several chunks of the mirror were knocked out by pistol shots. The pyrex did not crack, though, and the wire services reported that the astronomers had the telescope back in use the next night, finding its figure unaltered and its light-collecting capacity diminished only a few percent.

The 82-inch is now known as the Otto Struve Telescope. Recently run programs include image-tube spectrography of galaxies and photoelectric scans of nebulae.

In 1957 a 36-inch Cassegrain was added to the Observatory. This is used mainly for photoelectric photometry and spectroscopy; new methods of measuring, reducing, and interpreting data using auxiliary equipment such as a semiautomatic microdensitometer are constantly being developed.

As the exploration of the solar system by manned and unmanned vehicles got under way in the early 1960s, it became evident that there really was not much large telescopic equipment available for the most advanced study of the planets. Accordingly, NASA, the National Science Foundation, and the University of Texas contracted in 1964 to build a new large telescope at Mount Locke. This was to be used particularly for studies of planetary atmospheres, with high-dispersion spectrographic and spectrophotometric equipment which could add to what was known about the conditions at or near the surfaces of the planets. It was about this time that the Corning Glass Works developed the technique of making fused-silica discs of the sort used in the Naval Observatory's 61-inch at Flagstaff (q.v.), and this material was chosen as the best available for a 107-inch mirror. The mounting of the 82-inch had been a two-pier cross-axis English arrangement; this does involve adding a counterweight (not needed in a split-ring, fork, or yoke mounting) but at this rather low latitude it adds stability (and is cheaper) to build such a mounting. Also, a coudé beam can be directed down the polar axis with only two mirrors, using a large hole in the declination axis. So the 107-inch was mounted this way too, with the polar axis constructed of a double bent cone. The primary mirror is figured at $f/3.9$, and with the help of a corrector could be used at its prime focus. Actually the Cassegrain

The 107-inch telescope. The bent-cone English mounting requires a huge counter-weight. (McDonald Observatory Photograph)

focus (with the optics figured as a Ritchey-Chrétien) is more likely to be used, with an f/8.9 ratio for widefield work, and also f/18. With the coudé arrangement, the focal length is almost 300 feet. The mirror and tube are completely cased in steel, giving the appearance of an enormous boiler or tank.

Detailed spectrographic scanning of the planets is still a major function of the 107-inch telescope. Also, laser-ranging experiments using retroreflectors, placed on the moon by both Americans and Russians, can judge the distance with an uncertainty of only inches. Other sorts of projects are run, though, including the observation of absorption lines in the interstellar medium, and emission-line spectroscopy of nebulae.

The 82-inch Otto Struve reflector. (McDonald Observatory Photograph)

Every sort of astronomical investigation is going on at the University of Texas; a series of brief descriptions of the projects occupies ten pages of the 1975 volume of the *Bulletin of the American Astronomical Society.*

Visiting: A fine approach from the east is to leave I-20 at Pecos and head southwest on SR 17. This takes you through canyons whose walls are jointed lava, along an old cattle-driving route. Just before Fort Davis, turn north on SR 118. If you are coming from the west, you pick up 118 at Kent. The 107-inch telescope is open for a self-guided, slide-illustrated tour Monday–Saturday from 9–5 and 1–5 Sundays and holidays. On weekdays there are lectures at 9 a.m. and 11 a.m., and more frequently on weekend afternoons. About 50,000 people visit the Observatory annually. There is an excellent book shop in the lobby of the 107-inch dome; the *Geologic and Historic Guide to the State Parks of Texas* is available and recommended, in addition to many books on astronomy. The 82-inch telescope is available for public viewing on the last Wednesday night of each month; write McDonald Observatory, Fort Davis, Texas 79734 at least three weeks in advance, enclose a stamped, self-addressed envelope, and state how many will be in the party. Admission is by permit so obtained. This is by far the largest telescope in the world that is available at stated times for public observation.

Accommodations: There are strings of motels, some of which had rooms priced at $6 early in 1975, during the off-season, in Van Horn, Pecos, and Alpine. Between Fort Davis and the Observatory is the Davis Mountains State Park, which has a lodge with rooms for rent as well as sites for tents and trailers. This is a safe camping area with ideal sky conditions most of the time for the amateur astronomer. Vacation crowds may cause problems, but the park covers three square miles, and is only a thousand feet or so lower than the Observatory.

Nearby: Sunspot, New Mexico, with the Sacramento Peak Observatory, is a half-day's drive northwest; this is one of the great solar observatories, with a fine visitors' program. At the end of a gravel road just west of Odessa, Texas, is the second-largest meteor crater in the country; it is perhaps worth seeing to say you have been there, but in no way compares to the Barringer Crater in Arizona. In early 1975, deserted, and its small museum closed and covered with threatening signs, the crater looked a shooting gallery for beer cans, set up in an abandoned basement. McDonald is close to Big Bend National Park; one can buy maps and books about it and its geology at the Observatory book store.

Sacramento Peak Observatory
Sunspot, New Mexico 88349

On a peak in the Lincoln National Forest in southern New Mexico, there is an observatory whose sole purpose is to observe "an undistinguished middle-aged sample of the most numerous class of stars, about midway in the stellar ranges of age, brightness, mass, size, and temperature." The mediocrity here described is, of course, our sun; the language comes from an ingratiating section of the Air Force Cambridge Research Laboratory's *Report on Research*, about the activities of Sacramento Peak Observatory, the operation of which was taken over by the National Science Foundation in 1976.

The most spectacular instrument of the many located here is the 136-foot tower telescope, rising above the evergreens at its base like an Egyptian stele or obelisk. At the top, sunlight enters a 30-inch-diameter window and is reflected by two diagonal mirrors downward into one of the largest vacuum chambers on earth: it is 4 feet in diameter and reaches another 185 feet into the ground beneath the tower. Removing the air prevents currents that would distort the images of the sun, just as heat waves along the ground do on a hot day. The arrangement at the top of the tower is not quite the same as at Mount

Tower solar telescope. The rotating sphere with a quartz window is visible at the top. (Photo: Sacramento Peak Observatory, Air Force Cambridge Laboratories)

Wilson, where a heliostat sends the light downward; instead, there is an alt-azimuth turret, whose fused-quartz window is fixed in a 52-inch sphere that rotates to achieve altitude pointing. For horizontal movement, the entire telescope assembly, together with its observing floor and auxiliary instruments below ground, rotates on a mercury bearing at the top of the tower. This optical assembly weighs more than 250 tons—a mass equal to that of several heavy military tanks—yet small motors turn it. Tracking the sun by motions both in altitude and azimuth seems to present no problems; a small mirror intercepts part of the main beam and sends it through an optical system that produces a 3-inch image of the sun. As this image moves, photocells send signals directing the

Inside the Big Dome, with its unusual conical rotating roof. The solar spar aims several instruments at the sun at once. (Photo: Sacramento Park Observatory, Air Force Cambridge Research Laboratories)

servomotors. The motors operate constantly, and adjust the position of the turret without backlash. The height of the tower keeps the window above most thermal disturbances (for another method, see the McMath telescope at Kitt Peak). Photographs made here have shown details as small as 0.25 seconds of arc, the best definition of the sun's surface ever obtained. The 64-inch mirror at the bottom of the shaft, with its 180-foot focal length, can be tipped slightly to send the converging beam to any one of five exit ports. Outside of these ports are several different instruments for analyzing the sun, including spectrographs mounted in three vertical 5-by-65-foot vacuum chambers.

One special feature of a visit to this telescope is the opportunity to observe a closed-circuit live TV picture of the sun (taken through another telescope at the Hilltop Dome, about a hundred yards away).

Three pieces of auxiliary equipment, used with the Tower Vacuum Telescope, are of particular interest. The 12-meter Fastie-Ebert echelle spectrograph records spectra, using a grating to disperse a portion of the solar spectrum; that portion has been selected by a prism spectrometer. Actually, widely separated bands of the spectrum can be recorded at the same time, while three cameras photograph the sun's image on the slit of the spec-

The definition of the vacuum solar telescope is apparent here even to the untrained eye. Note the fine spicules, thought to be connected with the vertical magnetic field, made of jets of gas. In the lower center is an especially active region; a dark absorption area is seen in the upper part of the loops here. The photo was taken in H-A light. (Photo: Sacramento Peak Observatory, Air Force Cambridge Research Laboratories)

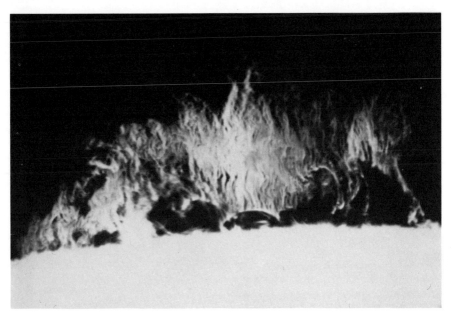

A small limb prominence about 30,000 miles high. It remains static while material flows down through magnetically defined "tubes". Photo was taken in H-A light. (Photo: Sacramento Peak Observatory, Air Force Cambridge Research Laboratories)

trograph; this provides a permanent record of what feature was being examined. It is possible to stretch out a 1-angstrom section of the spectrum to 16 mm at the highest dispersion. The AFCRL claims that this instrument has the best gratings ever made, and is therefore the best solar spectrograph in existence. A second important set of ancillary equipment is the system of narrow-band birefringent filters and 35 mm and 70 mm cameras that photograph surface features of the sun in fine detail; and indeed until you have seen some of these photographs, you have no conception of what a really sharp photograph of the sun is. One wonders why these are not included in more textbooks. The third important auxiliary is the multichannel magnetograph, which should analyze and plot the sun's magnetic contours with better speed and definition than earlier instruments.

From the visitors' gallery in the Big Dome, you can view a number of solar telescopes supported on a single spar. These include the largest coronograph outside Russia (a special telescope with a complicated set of lenses and a central disk that blocks the image of the sun so that its tenuous outer atmosphere, or corona, can be studied). The aperture of the coronograph is 16 inches, as is that of another telescope on the spar: the one that feeds the Doppler Zeeman Analyzer. The Zeeman effect is the splitting of spectroscopic lines in the presence of a strong magnetic field, and the purpose of the instrument is to plot magnetic contours on the sun. A third telescope on the spar is designed for photoelectric mapping of the outer reaches of the corona, and the fourth solar telescope here is a camera with a 15-inch lens that uses a very narrow bandpass filter to photograph phenomena at the extreme solar limb (the edge of the sun). There is another telescope in a roll-off shed just south of Big Dome; this is a horizontal long-focus refractor with a 12-inch lens that projects an image 36 feet into the observing laboratory below the Big Dome. Also in this laboratory are the spectrographs and the spectroheliograph that are bolted into the bedrock of the mountain.

Hilltop Dome contains another solar spar which carries patrol telescopes that record the image of the sun in hydrogen-alpha and white light; these operate automatically when the sun is out. One takes certain pleasure in referring to Grain Bin Dome; perhaps it ought to be renamed Sears Roebuck Grain Bin Observatory, since that is what it was, installed in the pioneering days of the Observatory. In use since 1950, it houses two small coronographs on a 10-foot spar; one of these records bright lines in the spectrum of the corona, while the other takes movies of the corona and by kinematic records has shown much about the dynamics of the corona as they relate to the sun's surface features.

As in any important observatory, there are quantities of analyzing instruments, and also supporting labs and shops. The Observatory operates a kind of space weather forecast, of great importance to the Air Force. The sun is largely responsible for all communications disturbances and for conditions in the ionosphere and above that affect the operation of satellites and missiles. Only an observatory of the kind at Sacramento Peak can adequately analyze the source of these problems.

Visiting: As with many other observatories in the Southwest, the view alone makes the visit worthwhile. At 9,200 feet altitude, you can look back to the Organ Mountains (that make the spectacular backdrop for the city of Las Cruces) and across the deserts to other mountains, a hundred miles and more away. The Observatory is open every day, for self-guided tours, from 9 a.m. to 4 p.m. In rare periods of great forest-fire danger it might possibly be closed. The booklet designed for this tour is the best that I know of; if you plan a trip west you might write in advance and request a copy. From May through October there are regular guided tours at 2 p.m. on Saturday. Sunspot is on most road maps, but in case you cannot find it, the community is located 18 miles south of Cloudcroft, New Mexico. The turn is a few miles east of Alamogordo, on US 82.

Accommodations: Extensive camping areas are found in the surrounding Lincoln National Forest, with a number of campgrounds located near Cloudcroft. In Alamogordo, White Sands Boulevard heads north from the center of town; this is also US 54, 70, and 82. There are a number of inexpensive ($7–8 single) motels, especially as you get into the 2400–2600 block. Ace Motel is one of these.

Nearby: Not many of the important observatories in New Mexico have regular visitors' programs. The astronomical traveler will scarcely fail to notice, however, that he is halfway between Kitt Peak and McDonald Observatory. Anyone headed north might check the phone number provided for the Langmuir Observatory to see if a visit is possible, and a detour to the site of the VLA is very much worth your while.

Kitt Peak National Observatory
Tucson, Arizona 85721

No photograph of Kitt Peak does justice to its natural setting. At 7,000 feet, out in the Sonora Desert 50 miles southwest of Tucson, the air is too clear: through the camera lens the perspectives are foreshortened and the mountains a hundred miles into Mexico look like nearby hills. The plains that are really thousands of feet below seem nearly level with the upper slopes of the mountain. It is only when you feel the wind that threatens to take you over the edge of the granite cliffs, and watch the ravens swooping and tumbling beneath, that you begin to understand the dimensions of the place.

From Tucson you head out State Highway 86, and to get onto this you have to drive to the south edge of town; the freeway connects with it. You actually only drive about halfway to Sells, and at a plainly marked Kitt Peak turn you head up the road whose grade, curves, and precipices compare well with Mesa Verde or Mount Wilson. Allow two hours to get there from Tucson.

51

Astronomers have learned to be very careful about site selection. Before the National Observatory began to be built here, more than 150 locations were considered and many were tested with telescopes hauled to their summits, on mules if necessary. The variables include weather (think of the problems at Yerkes), city lights (blinding Mount Wilson), vegetation cover, enough space for many instruments, and a supporting university. Kitt Peak approaches perfection by nature and by agreement with the city of Tucson and the county surrounding it, both of which require reflectors on area lighting to keep as much light as possible out of the sky. A camera sometimes monitors the skyline towards Tucson, looking for disturbing bright spots; these are located and subdued with reflectors.

But to the dismay of the men who had found such a perfect site, the Papago Indians (whose reservation you enter 20 miles before you reach the Observatory) were very unwilling to have their quiet, and especially the peace of mind of their deities, disturbed by programs that they imagined would involve shooting off rockets: a reasonable fear for anyone living a half-day's drive from the first atomic test. The astronomers took a group of tribal elders to the Steward Observatory on the campus of the University of Arizona in Tucson and showed them the moon and planets through the telescope there. One of them remarked that the surface of the moon looked much like their reservation. Anyway, they decided that the enterprise would not create much disturbance and an agreement with several interesting provisions was drawn up. The lease runs perpetually, as long as scientific work is taking place. The caves around the summit of the mountain are not to be entered, since their Papago god, Ee-Ee-Toy, might be occupying one of them, having taken the shape of one or another of the mountain creatures, as it pleased him. Ee-Ee-Toy ordinarily lives on the thumblike Baboquivari Peak that one sees 12 miles to the south: this, to the Papago Indians, is the center of the universe. Another lease provision is of great advantage to the visitor: Papago crafts are sold at the shop in the museum, and since the proceeds go directly to the craftsmen, there is no middleman. Not only are the prices far lower than elsewhere, the workmanship is superior. I brought home for my wife a basket whose green coils are unbleached yucca strips laced together in a set of concentric spirals of bleached white yucca or willow shoots. It is very solid and has the fragrance of desert vegetation. An attached card states that Susie Anselmo of Mais Vaya made this particular basket.

Many Papagos helped with the construction of the Observatory and many still work there.

What an aerial photograph of Kitt Peak does show is that the top of the mountain is littered with all size and manner of telescopes. The dome of the Mayall 158-inch reflector (third in the world in size after Palomar and the new Russian 6-meter) dominates the others, perched on top of the trussed cylinder that contains several floors of laboratories, offices, and storage space. The mirror is fused quartz; the section that was removed to permit the Cassegrain configuration sits in the middle of the museum. The bubbly murkiness of the glass reminds you that the purpose of the material is to reduce thermal expansion and not to transmit light. The limited flexure of the mirror gives it greater resolving power than Palomar's pyrex; the twin to the Mayall that is going up in Chile will be better yet, since its mirror is Cervit, a ceramic-vitreous material, with about a zero coefficient of expansion. Dark skies permit the use of the Kitt Peak instrument at the limits of the known universe. Its career is barely begun, as readers of *Sky and Telescope* (January, 1973) or *Arizona Highways* (November, 1973) will remember.

A large area near the Mayall telescope contains several instruments removed here from the University of Arizona, including not only the 36-inch reflector that the Papagos looked through but even the very large 90-inch. None of these is open to the public, but the University has visitors' nights in Tucson (see Steward Observatory).

There are about a dozen other instruments, ranging down to the small telescope off on a side ridge that from its silolike tower focuses permanently on Polaris in order to calibrate the seeing conditions. The National Radio Astronomy Observatory maintains a 36-foot dish that picks up wavelengths ordinarily absorbed by water vapor. This is invisible in its housing.

The unique Kitt Peak trademark is the Robert R. McMath solar telescope—the square white tower with the slanting shaft of the same dimensions (tipped at the angle of the latitude) thrust down into the rock of the mountain. Most photographs make it look as if it is out in flat country; you could take off actually a few feet away and glide for miles in a sailplane. The vertical tower is eleven stories high; the shaft heads down like a twenty-story building tipped at 32 degrees, but with a basement of another thirty stories—a 500-foot light path for the 80-inch wide beam of sunlight directed into it by the heliostat. A sort of funicular railway holds the mirrors inside. The main objective is a 60-inch paraboloid that sends a converging beam about two-thirds of the way back up the shaft to a 48-inch diagonal flat. The sun finally comes to focus on a horizontal table in the observation room below ground; its image is 34 inches across, and its intensity must not be a great deal more than that of ordinary sunlight. The difference is, of course, that here is a huge, sharp, focused *image*. Two smaller heliostats send their beams down the same tube so that multiple studies can be made simultaneously—of the spectrum and magnetic fields as well as the structure of the surface. Auxiliary equipment permits observation of the corona and of flares.

The solar telescope is designed expressly to overcome the problems of heating and air-current convection that accompany observations of the sun. The Big Bear Lake

The McMath solar telescope at the edge of Kitt Peak; the slanting section continues deep into the rock. Note the tubes for coolant bonded to the walls and the heliostat at the top. (Kitt Peak National Observatory Photograph)

Dome of the Mayall telescope, with Steward Observatory telescopes to the left. To the right is the small tower for the Polaris monitor and the dome of the NRAO millimeter-wave antenna. (Kitt Peak National Observatory Photograph)

solar telescope uses natural cooling by being located on an island in a lake; the McMath uses artificial cooling. As long as all the air inside the slanting square tube is kept ten degrees cooler than the air outside, it will sit there like a jug of cold water, without convection and without air currents. The thousand-foot path of the sun's rays will traverse a motionless medium. To achieve this, tens of thousands of gallons of chilled antifreeze circulate through copper tubing that is bonded within the copper sheathing that makes the walls of the telescope. The upper end is open, but the air settles and does not circulate.

There is a glassed-in visitors' gallery at ground level—about halfway to the bottom. When I was there a few cirrus clouds prevented use of the instrument, and we were allowed out on a catwalk in the middle where you could get a clear look 300 feet to the bottom. The mixture of acrophobia and disorientation (the catwalk being horizontal and everything else tilted) was superior to any fun house; I have never before become actually dizzy in a static situation, as my semicircular canals argued with my visual horizon.

Next to the McMath Telescope is a tower with clam-shell dome on top that holds a vacuum solar instrument. This avoids the distortion of air currents by evacuating the focal path.

Another important instrument—the largest, at first—is the 84-inch reflector which also has a visitors' gallery. It is in constant use, adapted to various programs with changing instrumentation. In 1970 this telescope recorded the spectrum of the most distant known object—or at least, the one with the largest red-shift—a quasar at 10 billion light years' distance. (I remember when this discovery put the last chapter of the astronomy text that I was using out of date and made one question obsolete on a final exam that I had prepared.)

Visiting: Kitt Peak is open every day except Christmas from 10 a.m. to 4 p.m. and there is no fee or ticket required. The best days for individual visitors are Saturdays, Sundays, and holidays since there is an hour-and-a-half tour at 10:30 a.m. and again at 1:30 p.m. The early hour is less crowded. On weekdays, tours for groups of twelve or more can be arranged in advance with the Observatory's Public Information Office. The tour is thorough and informative, though pitched a little low for advanced amateur astronomers. It starts with a twenty-five-minute film which includes an extremely interesting sequence on the grinding and polishing of the 158-inch mirror. One can hardly visualize these processes if one has not seen them presented kinetically. The conducted

The Mayall 4-meter (158-inch) telescope. It uses a horseshoe-shaped main bearing with oil pads at the bottom; the petals of the mirror covers are open. (Kitt Peak National Observatory Photograph)

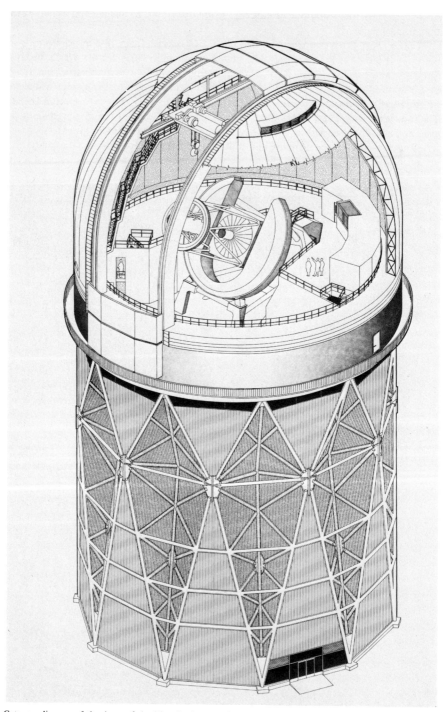

Cutaway diagram of the dome of the Mayall telescope; dark areas just below the catwalk are plate-glass windows for the visitors' promenade. The lower structure holds offices and shops. (Courtesy Kitt Peak National Observatory)

portion of the tour includes the solar telescope and the 84-inch. Go back to both after the tour is over and the viewing galleries are deserted. There is more to see alone, and actually the self-guiding tape in the McMath gallery had so much in it that I listened to it three times. Do not fail to walk up the hill to the 4-meter (158-inch) telescope. An elevator takes you up the last 75 feet of vertical rise, to the glassed-in panoramic viewing gallery. At least 200 miles towards every point of the compass! One appreciates the thoughtfulness towards the public in the inclusion of this promenade; and in fact one finds a friendly and informative welcome in the whole course of a visit. Inside, the 158-inch monster is illuminated about as dimly as Palomar's, but you can make out its essential parts—the vast split-ring mounting on its pressurized oil pad bearings, and the framework of trusses holding the secondary.

The museum in the visitors' center includes a working solar telescope with buttons by which the visitor can center the image in right ascension. Unfortunately, the children that come along prefer watching the sun sweep back and forth across the viewing area to studying any sunspots that may be visible; you wait your chance and get in the middle, between them and the controls. The displays are very well thought out. At the sales booth (in addition to the Papago Indian wares) you can buy color photos, postcards, books, and candy. There is a picnic ground about a mile down the side of the mountain on the road to the Observatory.

Accommodations: In Tucson there is a strip denominated The Miracle Mile (a real street name) that runs north from the middle of town, paralleling Interstate-10 with a few jigs and jogs thrown in. Coming from the north you take the I-10 business exit. Between the K-Mart at the north end of the Miracle Mile and the fast food places around where Speedway crosses it, there are about thirty motels of various sorts plus discount gas stations. Signs advertised rooms for $6 in February 1975, but for some this was a weekly rate and for others this seemed to apply to the first few empty rooms. Rates probably more than double when business booms. Twelve miles west of Tucson and on the way to Kitt Peak is the Palo Verde campground of the Tucson City Mountain Park. There is no camping on the Papago reservation. In the Coronado National Forest northeast of Tucson one finds the five campgrounds maintained by the forest service, up to 40 miles in the wrong direction. But if one could locate a site with adequate privacy and sky these might be excellent for using your own equipment. Seven miles east of Tucson off I-10 and a quarter mile north at the Craycroft exit is a private campground with more of the comforts of home.

Nearby: Anyone who has reached Tucson by car and who is interested in astronomy should go to Flagstaff. A day's round-trip drive is not out of the question, to reach the Lowell Observatory by 1:30 p.m. for an hour's visit, and for a half-hour look-in at the Naval Observatory 6 miles outside town (see descriptions of both Observatories). But it would probably be better to do it in a more leisurely way. A circular route could take one by Casa Grande National Monument, through Phoenix, up Oak Creek Canyon by alternate US 89 after a look at Montezuma Castle, and into Flagstaff by noon. A return another day by way of I-40 to the Barringer Meteor Crater and through the Petrified Forest would bring you back to Tucson on state road 77 through a lot of spectacular volcanic and canyon scenery. Tucson is also a good jumping-off or returning point for Mexico.

For a person staying in Tucson, a visit to Steward Observatory and the Flandreau Planetarium is obligatory; see their descriptions in the Arizona section of Part II. Mount Hopkins and the Catalina Mountain Complex are not ordinarily open for visits, but you may want to read about them. If you are headed east from Tucson, look into the possibility of visiting Sacramento Peak, in southern New Mexico.

Lowell Observatory
Flagstaff, Arizona 86001

The Lowell is one of the few major observatories that is easy to get to by public transportation. From Flagstaff's main street—where both bus and train terminals are located—you can see it up on the hill at the west end of town: the white, metal-sheathed wooden dome that Percival Lowell ordered built in Mexico as a temporary telescope cover. The cover was so satisfactory that he had it dismantled and set back up here. The northern Arizona pine forest, which reaches to the edge of Grand Canyon 60 miles to the north, covers the hillside.

That dome houses the only telescope ordinarily viewed by visitors: the 24-inch Alvan Clark refractor, first installed in 1896 and in use ever since. You may see it fitted with a movie camera that takes a frame every few seconds during clear nights. Together with other telescopes around the world, this one participates in the International Planetary Patrol, which keeps a 24-hour record of the surface features of planets, in case anything unusual should appear. The mesa on which the telescope is located raises it 300 feet above Flagstaff; often it is possible to take advantage of a stable thermal inversion pattern that not only makes for steady planetary images, but also takes a little of the chill off the night. But the astronomers also use electrically heated flight suits. Since Flagstaff is at 7,000 feet elevation, much of the earth's atmosphere is below it. At one time the dome used cannon balls for bearings, and boys were hired to come up from town at night to provide the power to rotate it. Now, even less elegantly, it turns on automobile tires bolted to the walls—some with hubcaps and some without—and, yes, they did once have a blowout.

But for anyone with a sense of history and a love of fine optics, the visit to this telescope can be more rewarding than any other. There is much satisfaction in recognizing that the interior of this dome, and the long riveted tube of the telescope, are very nearly the same as when Percival Lowell used it. He worked himself into a state of nervous exhaustion, peering for momentary glimpses of especially good "seeing" that would reveal more of the spiderwork pattern of "canals" on Mars. The fact that these were more the work of an overstrained imagination than of accurate observation should not be more disturbing than Kepler's nests of perfect polyhedrons circumscribing the planetary orbits.

Pluto, the first two letters of whose name memorialize Mr. Lowell, is, of course, the most celebrated discovery of the Observatory. Mr. Lowell had been dead fourteen years when, in 1930, a special wide-angle 13-inch astrographic telescope (much like an aerial camera) finally picked up the tiny moving dot of the outermost planet—close enough to Lowell's predicted position to allow him credit for the discovery. Recently some have questioned whether luck played a part in this; but the fact that Lowell employed a dozen mathematicians to carry on computations for ten years (on the orbital disturbances of Uranus and Neptune) proves that if indeed it was luck, no one did more to earn it.

Much more impressive to us now was the discovery here of the "red-shift" in the spectra of distant galaxies, first observed between 1912 and 1920. The cosmological implications are still being worked out. The immediate effect was to extend the borders of the known universe enormously, and to suggest that the universe was expanding.

If you visit or write to the Observatory, you will receive a pamphlet listing other important discoveries, research programs under way, and the many other instruments housed on the 700-acre Flagstaff domain, or a few miles away. These include the very large 72-inch Perkins Reflector used jointly by Lowell, Ohio State, and Ohio Wesleyan, and a number of large refractors and reflectors in the 20-to-24-inch range. These are

Percival Lowell's 24-inch Clark refractor; the plank structure of the dome is visible in this old photograph. The telescope is still managed partly with ropes. (Lowell Observatory Photograph)

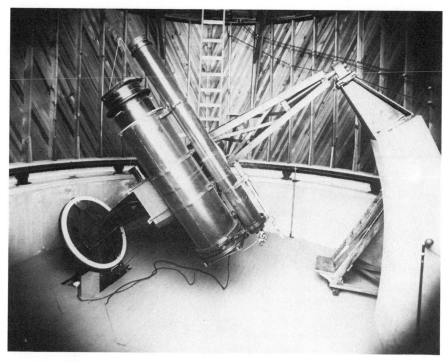

The 13-inch astrographic telescope designed especially to search for Pluto. It is now used for proper motions, comets, and minor planets; it resembles a huge camera. (Lowell Observatory Photograph)

not ordinarily shown to the public. If you wish to view a large reflector though, all you have to do is drive 6 miles to the Naval Observatory (q.v.).

Visiting: A talk, and a tour of the Clark refractor every weekday at 1:30. Not open Saturday, Sunday, or holidays. Sometimes you are met by a resident astronomer; perhaps more often by a graduate student. Your guide will in any case be well-informed about astronomy, and probably infected with an appreciation of Percival Lowell's personality. You should drive (or walk) up a little early, since at about 1 p.m. you may well find the door unlocked and ajar to the circular library. It contains many interesting displays and perhaps a few items, such as maps of the moon or Mars, for sale. Cases around the walls contain old instruments, meteorites, and so on. A good many of the books here were Lowell's own; the upper tier of shelves holds books of mathematical puzzles, which he liked.

During the summer, ask about visitors' nights, which may be held every other Friday. Free tickets may be obtained at the Observatory or from the Flagstaff Chamber of Commerce.

Accommodations: There is a solid string of motels that extends east along old U.S. 66 for about 2 miles from the middle of town. Some have off-season rates in the $6 to $8 range; these will fluctuate with demand. But remember that the highway parallels the Santa Fe railroad tracks; to escape the trains go on through town and under the underpass (or stop sooner on the way in from the west). In February, 1975, I had a nice room at the Flamingo Motor Hotel for $6.50, with a carport that kept off the snow that you do get here in the winter. For campgrounds, you can head to the ones called Kaibab Lake

The 72-inch Perkins reflector, used jointly by Lowell, Ohio Wesleyan, and Ohio State University. It rides on an English mounting with a heavy counterweight; note the immense setting circles, and the spectrograph attached at the Cassegrain focus. (Lowell Observatory Photograph)

and Parks in the Kaibab National Forest; east of Flagstaff is a whole set of commercial campgrounds and trailer parks. Actually the National Park campgrounds along the Grand Canyon are not all that far away.

Nearby: See descriptions of the Naval Observatory and the Meteor Crater. Between Flagstaff and Grand Canyon stretches the San Francisco Volcanic Field. The mounds along the road in this area are mostly what they look like: defunct cinder cone volcanoes. It looked to me as if what was being spread on the icy roads as I left Flagstaff in a 5 a.m. blizzard were these very cinders, mined from a quarry in the side of one of the volcanoes. For the most interesting part of the volcanic field, take the drive through the Wupatki and Sunset Crater National Monuments. The Indian ruins are interesting, and the fresh lava flows from this rather recent (700–800 years ago) volcanic episode give a feeling for dynamic earth processes. "People erosion"—too many hikers—has caused Sunset Crater itself to be placed off-limits; the Park Service is meditating on another mode of access to the crater.

The Museum of Northern Arizona in Flagstaff is worth visiting; the Museum and the Grand Canyon Natural History Association have recently published an excellent book: *Geology of the Grand Canyon.* Few areas in the world open up our planet on a dissecting table like this. If you are staying in Flagstaff you can make a feasible day's tour out of a drive to the South Rim; a couple of hours for stops at shops, museums, and scenic points; and a drive down recently improved S. R. 64 to Cameron. This not only takes you along nearly 20 miles of the Canyon's brink, but also permits views of the canyon of the Little Colorado. Your return takes you through the Wupatki and Sunset Crater Monuments. The Grand Canyon, of course, deserves a great deal more than this.

Other side trips can include Oak Creek Canyon to the south and possibly the Petrified Forest, 120 miles east and right past the Barringer Meteor Crater. A round-trip one-day drive to Kitt Peak would not be a good idea; Kitt Peak and Tucson are now the astronomical center of the world, and an adequate visit of that area demands a minimum of two days.

United States Naval Observatory—Flagstaff
Flagstaff, Arizona 86001

As useful as the observations of the Washington Naval Observatory may be, the location is surely one of the worst in the country. Taking advantage of Percival Lowell's much earlier surveys, the Navy chose a hilltop near Flagstaff to install its 40-inch Ritchey-Chrétien reflector in 1955. The air here at 7,600 feet is clear; light pollution from Flagstaff, though not welcome, is no more than that emanating from a small part of one suburb in Los Angeles; the urban area seems unlikely to expand very much; there are few sources of smoke or fumes; and local inversion patterns lead to especially steady "seeing."

Observations made with the 40-inch showed that the large number of clear, steady nights did make this a fine location, but the configuration of a Cassegrain telescope does not lend itself well to the Navy's favorite branch of astronomy, which is astrometry, or the measurement of star positions. Forty inches is plenty of aperture for studying comets and minor planets, and for registering the brightness and spectral types of faint stars. The problem is the convex secondary mirror, where a minuscule error of position can cause it to change the plate scale of a photograph slightly, or to shift the image of one star closer to that of another. For many purposes such tiny errors do not matter, but for astrometry the errors obscure the minute position shifts that the observer (actually, the person studying a photographic plate) is trying to measure. Very small errors of figure occur in any optical telescope, and after long experience with a refractor, those who use it learn to allow for these as part of the process of "reduction" of their observations. In a Cassegrain telescope there are two movable optical surfaces, and problems arise from their positioning with respect to one another.

In order to extend astrometrical observations to very faint stars, the Navy, early in the 1960s, tried out a new concept in telescopes. This is the 61-inch folded Newtonian that you are allowed to view (from a visitors' cage, unless, like me, you arrive alone in the middle of winter and are permitted out on the floor of the dome). Every feature of

The 61-inch folded Newtonian astrometric telescope. With an f/10 primary, it gives unusually sharp images. (Official U.S. Navy Photograph)

this instrument is aimed at achieving the predictability and reliability in the photography of star images that usually is limited to refractors. What you notice first is the massiveness of the fork mount, and the unusually thick struts that support and brace the secondary. Since it is impossible to eliminate all flexure in a trussed structure like this, the design allows the cells holding both the main mirror and the secondary to bend (from gravity) exactly the same amounts so that the alignment of the mirrors stays the same as the telescope is aimed at different parts of the sky. The 5-foot mirror, which weighs more than a ton, was intended to expand and contract as little as possible, and hence was made from fused quartz by a process that involved spraying a solution containing silica onto a revolving disc inside a refractory furnace. Today, the material of choice would no doubt be Cervit, which has an even lower coefficient of expansion. This mirror was ground and polished to a focal length of about 50 feet, giving a focal ratio of f/10; this is unusually large, three times that of Palomar, in fact. But whereas the 200-inch mirror serves mainly as a "light-trap" for the most distant objects, the 61-inch has as its purpose producing the sharpest possible individual star images. Furthermore, astrometrical measurements are all within our galaxy, mostly within our local region of the galaxy. Great photographic "speed" would be perhaps desirable for taking the position of some dim white dwarfs, but precision of star image is more important. A long focal ratio produces a large plate, which is free of the aberration endemic to reflectors known as "coma," or the smearing of star images at some distance from the center of the field. In fact the telescope has made some of the most beautiful, sharp photographs of galactic and globular clusters that I have ever seen; while this is not its purpose, one cannot help suspecting that those among the staff with the instincts of amateur photographers enjoy making these. They must look forward to trying an occasional "test," registering on film the symmetrical glory of a globular rather than another of the necessary, but shapeless, small star fields.

If it were practical, the ideal arrangement would be to place the photographic plate at the prime focus of the 61-inch. This, however, would mean putting a camera and possibly an observer at the other end of a 50-foot truss instead of at the base of a 25-foot one. The compromise is the flat mirror that serves as a secondary. Unlike the Cassegrain secondary, the only critical alignment here is that the mirror be precisely perpendicular to the optical axis of the main mirror. This flat secondary reflects the cone of light directly back through a hole cored out of the primary—so the configuration is really that of a "folded Newtonian," though the flat is proportionately more than twice as large as that in a regular Newtonian.

Two important types of astrometric work might be mentioned in connection with the 61-inch. In making the first step into the Universe from the Solar System, it is important to determine with great accuracy the distance to the nearest stars. The method of trigonometric parallax is as old as the Egyptian surveying techniques: you use a base line (in this case, the width of the earth's orbit around the sun); you photograph a star field twice at six-month intervals; you measure the angular displacement of a nearby star relative to a very distant star or (better) a galaxy; and you compute its distance using the familiar trigonometric functions that at least used to be taught in high school. The problem with this kind of "surveying" is the very tiny angles involved, and no such parallax was ever successfully measured until about a hundred years ago. When the distances of the nearer stars are known, their types and apparent brightness provide measuring sticks to reach even further into space. A second important type of astrometry is that of the "proper motions" of the stars. Again, we are usually talking about nearby stars, whose positions seem to change ever so slightly against a background of more distant stars and much further galaxies. This occurs mainly because of the rotation of our galaxy, and also because of the impetus that may have been lent a star as it condensed in a swirling

This photograph of the M 13 globular cluster in Hercules shows the wide field and sharp images of the 61-inch; a very slight guiding error is present, but no sign of coma at the field edges. (U.S. Naval Observatory Photograph)

cloud of hydrogen. The most important result of such measurements is to determine the motion of the galaxy and the motion of the solar system relative to other stars, within that general rotation.

The 61-inch seems to be fulfilling its purpose admirably, though there are those, at Yerkes, for example, whose love for their fine old refractors reinforces their conviction that the rigidity of the refracting lens cannot be equalled by any combination of mirrors.

Visiting: It would not hurt to give the Observatory a call from Flagstaff before driving out the 6 miles (you could walk out, even, if you came in on a bus). If you arrive on a week-day between 9 and 11 a.m. or between 1 and 4 p.m. there will probably be someone to show you around, or at least let you into the visitors' gallery. Head out of Flagstaff on Alternate U.S. 89 and drive under I-40; after a mile or two start watching for the turn-off to the right, which is plainly labeled. The Observatory is about a mile up the road.

Accommodations, etc.: See the description of Lowell Observatory, which you certainly must visit if you are in Flagstaff.

The Hale Observatories:
Mount Wilson and Palomar Mountain
Pasadena, California 91101

Here are two observatories that have caught the public imagination in this century as no others have. Probably this is true because between them they have been able to claim ownership of the largest telescopes ever made, with Wilson's 100-inch holding the title from 1919 until Palomar took over in 1947. And until word comes that the Russian 6-meter reflector is at work, Palomar can continue to hang on to its laurels.

The Hale Observatories are operated jointly by the Carnegie Institution of Washington, and the California Institute of Technology in Pasadena. Both Observatories owe much to the efforts of George Ellery Hale (as did Yerkes) and both belong to the same episode of great mirror-making. Also, it happens, visits to both will be similar; the pattern of self-guided tours, limited access for the public, "museums" of enlarged transparencies, and glassed-in visitors' galleries, indicates identical planning at each installation. The ordinary visitor, it must be said, may leave with a slight sense of disappointment, especially if he knows a good bit about astronomical equipment and would like to learn more. Because the lights are kept low, and the visitors' booths are glassed in (both measures necessary to keep the temperature down), the view of the giant telescopes is like that enjoyed by those who saw the Titanic founder: impressive but dim. The observatories could easily be equipped with a *son et lumière* set-up, whereby low-voltage spotlights (that would add only a few watts to the thermal contamination of the dome) could be coordinated with a taped lecture. Kitt Peak has an excellent push-button lecture at the McMath Solar Telescope, and so does McDonald. In view of the hundreds of thousands of visitors per year, the outlay would be small indeed: ten automobile headlights, a tape recorder with a cue switch, a cheap transformer, and a few hundred feet of wire would do it.

MOUNT WILSON. The road up Mount Wilson, especially the last 5 miles (the next-best thing to a Grand Canyon mule trip), brings to mind the early years of the Observatory. One should imagine the curious vehicle—run by a gasoline-powered generator that delivered electricity to the four individual wheel motors—that delivered and retrieved astronomers and materials from Pasadena below. Old photographs of the road have scenes with titles such as "The Hundred-Inch Has a Close Call," showing mules, men, cables, and winches about to slip over a precipice.

Once you get to the top, you can look off Signal Point (so named because old-timers lit a fire to show that they had got there) in the direction of Los Angeles. If there is not too much smog below, you can see enough of the city to know why in the long run Mount Wilson has not been a good bet as an observatory site. Fortunately, the astronomers were so industrious that, before lights blotted out dimmer objects, they almost made up for this particular short-sightedness. Up here on top, the peaks of the San Gabriel Mountains collect enough moisture for a good cover of pine and fir; such vegetation is always valued as a means of cutting down on thermal air currents, and has been especially helpful for the solar work. There are also thickets of the evergreen oak, called

Aerial view of Mount Wilson and the San Gabriel Mountains. From left to right are the two solar towers, the dome of the 60-inch, and the Hooker telescope dome, plus other small structures. (Hale Observatories Photograph)

chaparral, although this plant usually flourishes at lower elevations than the Observatory's 5,700 feet. Along the highway you may have already noticed the desert yucca called "The Lord's Candle;" it produces a huge flowering stalk and then dies—unlike other yuccas. The drive up and the exposures of rock around the summit are also geologically instructive. Mount Wilson is actually a lower peak in a system of igneous and metamorphic rocks which are in part a "horst," or a block of crustal material elevated between two faults.

The most powerful optical instrument on the mountain is the 100-inch Hooker telescope. Work on this began in 1908 after the completion and installation of a 60-inch reflector; the optician was George Willis Ritchey, co-inventor later of the Ritchey-Chrétien telescope. At that time large glass discs were still imported from France; when the 100-inch slab arrived, it contained layers of bubbles where the glass had been ladled in. For a year it was set aside as unusable, and even when grinding began, Ritchey thought that the bubbles would cause uneven expansion of the mirror (with temperature changes) and make it worthless. He was so pessimistic that, after many years of grinding and figuring, and successful optical tests in Pasadena, he refused to help with the war-delayed installation of the mirror in 1917. For a few hours it seemed he had been right, but then the mirror "settled down" as it reached an even temperature. The mounting of the telescope has a desperate look to it; all girders and rivets, and so ready to compromise with the requirements of rigidity that a large circle around the celestial pole is inaccessible. It is suspended in a rectangular yoke that rotates on mercury bearings. An illuminated chart just outside the visitors' gallery illustrates the various focal arrange-

The 100-inch Hooker telescope; the yoke mounting turns on mercury bearings. (Hale Observatories Photograph)

ments. Beside the door of the dome (which has double-walls to help even out temperature changes for the sensitive mirror) is a section designed to be bolted to the upper end of the tube, to hold a secondary; one can get a feel for the size of the telescope by standing inside the bars and struts of this piece of framework.

A great deal of basic astronomical work, especially that having to do with accumulating the spectra of stars and galaxies, has been carried out with the 60-inch and the 100-inch reflectors. Determinations of radial velocities (and hence the size of the galaxy and the universe, as well as their shapes and motions) are an important part of this, by the familiar method of studying Doppler shifts in the spectral lines. Directly and indirectly such studies also produce most of the information that we have about the constituents, temperature, structure, and even size and weight of stars.

Another conspicuous instrument at Mount Wilson is the 150-foot solar telescope on top of the latticed framework, with a remote-control dome. The upper platform holds a coelostat (a mirror that tracks the sun) which directs a beam downwards, where it is focused by a lens and directed into a well-like observing laboratory dug into the ground below. The vertical arrangement cuts down distortion from currents of air heated at ground level; the great length gives a large solar image for study and photography, and a widely dispersed spectrum for detailed resolution when the 75-foot spectrograph is used.

One unusual program at the Observatory was the development of interferometry to measure the diameters of larger nearby stars (red giants). By this technique, invented by Michelson, mirror flats separated by a much wider distance than is feasible as a mirror diameter (50 feet) direct starlight into a single telescope. The pattern of lines where the light interferes or reinforces itself gives a measurement of the size of the object.

The 150-foot tower telescope. The clam-shell shutter is open, letting the sun shine on the heliostat that sends its rays down the enclosed tube; the tower helps to avoid distortion from ground heat waves. (Hale Observatories Photograph)

Although the Michelson interferometer has not been used since 1933, at radio wavelengths this technique now employs the whole surface of the earth as a baseline (see Green Bank).

Each of the major telescopes mentioned above is equipped with various kinds of auxiliary photometric or spectrometric instruments; over the years, just as in all active observatories, these instruments have been constantly improved, using new developments in optics and electronics. The coudé spectrograph of the 100-inch telescope is perhaps the most important of these: used with an image tube, it has recently helped in the search for optical equivalents of X-ray sources.

Other programs recently under way at Mount Wilson include a continuation of solar observations. The day that I visited, the 150-foot tower telescope seemed to be open and functioning, despite some high clouds. The solar telescopes operate daily, providing photographs, spectroheliograms, magnetograms, integrated-light magnetic-field measurements, and sunspot drawings. Planetary studies are not neglected; an interferometric scanner at the coudé focus provides information on the atmosphere of Jupiter and Titan (Saturn's largest satellite). Also, the 100-inch is used successfully for infrared photometry of faint infrared stars in the Orion Nebula—the sort of investigation that may fill one gap in the earlier stages of the evolution of stars, when the hydrogen is just approaching the thermonuclear ignition point. Even the 60-inch and a smaller 40-inch reflector figure prominently in reports of research carried out using new types of narrow-band filters and specialized photoelectric detectors.

In addition to the large instruments, which continue most in demand for research projects, there are other smaller reflectors, refractors, and cameras housed in smaller domes around the mountain.

PALOMAR MOUNTAIN. Separated from Los Angeles by both miles and mountains, Palomar continues to enjoy relatively dark skies. The drive south, if you are taking US 395 and then SR 76, leads you through stretches of citrus groves, past banks where flowers bloom in what is the dead of winter in most of the country, and finally up into valleys that are dry enough to make prickly-pear hedges possible around some of the houses. The higher slopes of the mountain itself are heavily forested with evergreens. As you reach the top, after passing the two Forest Service Campgrounds (now on Road S6) you may wish to stop at the picnic ground and take a look at the exposed rock in the road cut. This shows what the mountain is made of, a much-invaded granitic mass that rides, a solid block, between the Elsinore and Agua Caliente faults that separate it from the sand and gravel plains to the east and west. This 10-by-30-mile rock island may preserve the telescopes from major damage in case the San Andreas fault lets loose again. The section in the bank across from the picnic area gives instant proof of at least three geological eras: a dark basaltic dike cut across by veins of quartz that also traverse the surrounding matrix.

As at Mount Wilson, the dim view from the visitors' gallery, however necessary for the protection of the telescope, may be a little disappointing, and the museum is rudimentary. In summer the grounds may be swarming; the two observatories report 250,000 visitors a year. But the drive and the views of the California coastal ranges would be worth it in themselves. Furthermore, one does carry away a somewhat improved impression of the scale of the Hale Telescope—which, because of the squat massiveness of the split-ring mounting, actually almost looks more compact and manageable than some of the largest refractors. Although the objective is five times the diameter of the Yerkes lens, the main tube is in fact 20 feet shorter; adding to the impression of squatness is the fact that the tube section (44 feet long) rests in a cradle that is even wider—46 feet across. The cradle alone weighs 340,000 pounds.

The pouring and grinding of the mirror for the Hale telescope—whose completion was melodramatically delayed for ten years during World War II—captured the popular imagination completely. The mirror blank's progress across the country (on a special flat car, whose lowest part cleared the roadbed by mere inches) was followed closer than that of visiting royalty. School children of that era heard about it constantly, and its completion inspired the writing of many books. As early as 1934, the first efforts in the casting of the disc took place at the Corning Glass Works. There was plenty of drama at this stage, with ceramic cores—that were supposed to make a honey-comb structure in the back—floating up to the surface; then the months of slow cooling and annealing that followed the second effort. Optical preparation was carried out years later, after the blank had crossed the country to the workshops at Caltech. Finally, in 1947, the finished mirror made its way like a juggernaut through crowds of astro-nomical worshippers, south from Pasadena and up the last steep, winding miles.

The dome of the Palomar 200-inch in the moonlight. The split-ring bearing is visible through the slit. (Hale Observatories Photograph)

Palomar's 48-inch Schmidt. The correcting plate is 2 feet less than the 6-foot mirror at the back, so the tube tapers; an observer demonstrates the guidescope. (Hale Observatories Photograph)

These are some of the most interesting parts of the completed Hale Telescope:

The Main Mirror has a 40-inch hole in the center for the Cassegrain focus, and is shaped as a paraboloid with a 660-inch focal length. It can be removed by a crane and lowered into a vacuum chamber when it is necessary to replace the aluminum coating. It is supported from the back by thirty-six complex devices that are inserted in holes; each has more than a thousand parts, and compensates for changing loads as the position of the mirror changes. This keeps the shape of the mirror stable.

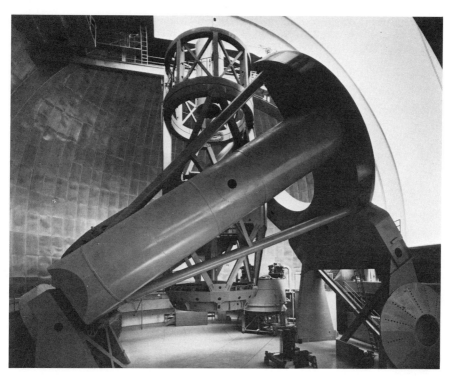

The 200-inch Hale telescope. Note the dolly on rails, at lower center, to transport the mirror for re-aluminizing. (Hale Observatories Photograph)

The Observer's Cage is located at the other end of the tube; an elevator in a circular track on the side of the dome provides access to the cage. If the telescope is used at its prime focus, a correcting plate is interposed in the converging beam to extend the field of good definition; the observer loads the film holder and guides the telescope with a microscope which is set on a guide star at the edge of the field. On the bottom of the observer's cage are hyperboloid convex mirrors that can change the focal ratio to f/16 or f/30; the first ratio sends the beam through the hole in the main mirror, or, by means of a flat, out the east arm of the yoke; the longer ratio is for the coudé focus below the south polar bearing. Several flat mirrors can be mounted or interposed in various ways within the tube.

The Yoke is in the shape of a "split ring"; this makes it possible to tilt the telescope directly toward the north celestial pole. The horseshoe at the north end uses "oil-pad" bearings: a pump forces oil out through holes so that the smooth horseshoe actually floats on an oil film; scrapers remove excess oil as the telescope turns, and the oil flows back to a reservoir. Friction is so low that a 1/12-horsepower motor drives the telescope when it is tracking. For fast slewing, to acquire an object in the sky, a 2-horsepower motor suffices. Ordinary roller or ball bearings would have required five hundred times the power. Oil pad bearings also serve the south end of the yoke, but in this case they are shaped hemispherically to take the thrust as well as the lateral load. These bearings were so successful that they made the telescope sensitive to wind-induced oscillations; when it is not slewing, therefore, a friction device makes it a little harder to turn, and thus damps the vibrations.

73

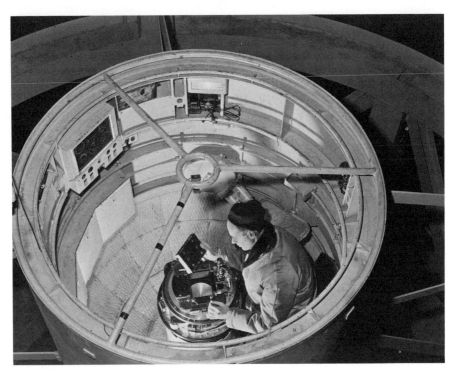

The observer's cage of the 200-inch; a photographic plate and the eyepiece for guiding can be seen. (Hale Observatories Photograph)

The Dome is about 135 feet in diameter—yet the telescope seems so nearly to fill it that one can hardly believe that a thirteen-story building might fit inside. The dome weighs a thousand tons, being constructed with an outside shell of steel plates and an inside layer of boxlike structures filled with crumpled aluminum foil. A 4-foot separation provides ample space for air circulation to keep it cool in the daytime. At one time, the geometrically insoluble problem of coordinating dome rotation with telescope movement was solved by using a small model to activate motors that drive the dome. When contacts on a little dome encountered those on the model telescope, the actual dome moved (followed with synchro motors on the model). Now electronic chips can carry out the calculations necessary to match the movements of the telescope and its cover.

New auxiliary equipment for the 200-inch telescope is continually under development. One may mention the multichannel spectrometer, the image tube for direct photography, the SIT-Vidicon digital photometer, the digital data system, and other advanced developments for data acquisition and telescope control.

New projects are run on the 200-inch as fast as competing astronomers can think them up and get approval. Simultaneous observations of the Cygnus X-3 X-ray source have been made with the Copernicus Satellite and the Hale Telescope, which was operating at the 2.2-micron infrared wavelength: synchronous short-term variations proved that both were observing the same object. A new device—the indium-antimonide detector—has been used to observe Jupiter in the infrared. Millimeter-wave observations made at twilight turn the big paraboloid into a high-resolution radio antenna. Image-tube spectra provide information on the rotation of galaxies. It is a pleasure to learn that this instrument is still operating on the cosmological frontier; using a 90-mm magnetically focused image tube, plates are taken at the prime focus (after having been baked to increase sensitivity) in a systematic survey of the faintest possible objects.

Several hundred yards to the east of the Hale Telescope is a smaller dome that houses the 48-inch Schmidt, one of the largest such wide-field cameras in the world. The 4-foot aperture is determined by the size of the correcting plate; the mirror at the rear is actually 6 feet across. A curved holder for photographic plates is located at the halfway point of the 30-foot tube. For many years after its installation, the main work of this telescope was the Palomar Sky Survey. This was completed in 1959, and copies of these plates and prints thereof are used worldwide for research and discovery. Nowadays the Schmidt is under heavy demand for research projects. These include searches

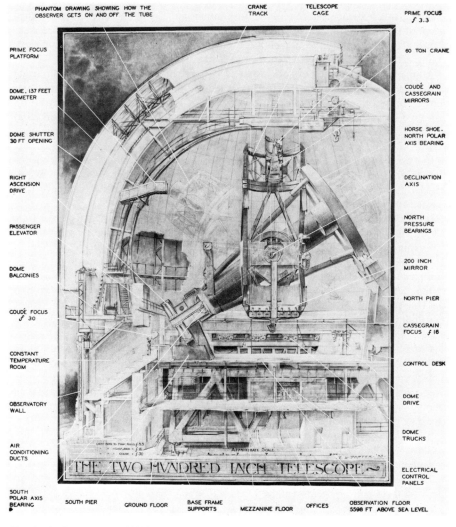

Drawing by Russell Porter of 200-inch telescope. (Hale Observatories Photograph)

for optical counterparts to radio sources, and surveys designed to turn up supernova explosions. Also, galactic evolution may be explained by studies of interacting galaxies made with the Schmidt; spiral arms may come into existence because of tidal disturbance of a galaxy.

In 1970, a 60-inch reflector went into service at Palomar, and this is inside the dome located northwest of the Hale Telescope. The main purpose of this was to provide a large-aperture instrument for photometric work and for direct photography. Mount Wilson's increase in sky brightness made such a telescope especially desirable for work on faint objects.

The telescope has a flip-flop secondary mirror, which also moves along the optical axis for focusing. In the Cassegrain configuration (Ritchey-Chrétien system) it operates at $f/8.75$, and at $f/30$ for the coudé focus. A flip-out mirror sends the coudé beam out through the declination axis, and two other flats direct it back down the polar axis. The telescope can therefore also be used for high-dispersion spectrography. Like the 48-inch Schmidt, this telescope rides in a large fork mounting. The mirror uses the fused-silica material that was the most temperature-insensitive substance available until the invention of Cervit. Recently the 60-inch has been used in hydrogen-alpha surveys of interacting galaxies, and in spectrographic studies of galactic rotations.

All the instruments on Mount Wilson and Palomar are available to staff and outside astronomers, who submit proposals that are judged for merit by the Observatory Committee. In addition to their observational work, a number of staff members also do some work of theoretical nature. Furthermore, some theoretical work in astronomy and astrophysics is done at Caltech by professors who are not members of the Observatory staff.

In Pasadena the Hale Observatories have an electronics development laboratory, a photographic laboratory, a machine shop, one of the best optical shops in the world, measuring laboratories, plate collections, and computers. Libraries and the offices for astronomers, engineers, and administrators are also located there.

In 1963, with the cooperation of the University of Chile, astronomers from the Hale Observatories carried portable telescopes to various Chilean mountain sites to test observing conditions for a southern-hemisphere station. The peak at Las Campanas offered clear skies and atmospheric conditions of exceptionally good quality. In 1968, construction began for an observatory in a location 40 miles from the coast, at an elevation of almost 8,000 feet. Since 1971 a 40-inch telescope has been in operation here, and the University of Toronto has installed its own 24-inch. In the course of 1975, the new Du Pont telescope, with an aperture of 100 inches, was being assembled and tested at this site.

Visiting the Hale Observatories: To see Mount Wilson Observatory you must enter Mount Wilson Skyline Park, and pay a fee of $2.00 per carload or $.50 per pedestrian. This is open daily 9 a.m. to 5 p.m. from April to mid-September, and 10 a.m. to 4. p.m. Wednesday through Sunday in the winter. Once you get on SR 2 there is no problem finding the Mount Wilson Road, about 18 miles northeast of Pasadena; getting onto SR 2 can be a problem, because the exit signs call it "Glendale Boulevard" if you are on I-5, and "Canada Boulevard" if you are on the Colorado Freeway (134). Be prepared to get lost.

Palomar Mountain is open 9 a.m. to 5 p.m. all year. There are various ways of getting onto US 395 south of Los Angeles; from US 395 you turn east on SR 76 and continue until S6 takes you up the mountain. *Don't pull a trailer either place!*

Accommodations: For purposes of visiting the Observatories, San Bernardino makes a central location. Furthermore, there are several moderately-priced motels in the middle of San Bernardino. Take the Fifth Street exit off the US 395 freeway and go two blocks east. You will see the Imperial 400 and the Desert Inn motels, which had singles around $10 in 1975; if you continue to "D" Street and turn left, you will find—a few blocks north—the Civic Center Motel. I had a very nice, very quiet room here for $7 a night in February, 1975. For campers, there are numerous sites in the Angeles National Forest, and amateurs from Los Angeles often use these for dark-sky viewing. There are actually two Forest Service campgrounds within a mile or two of the Palomar Observatory; in warm weather it may be hard to find a vacant tent or trailer site, though.

Nearby: See the California section of Part II of this book; among the listings there are the Jet Propulsion Lab in Pasadena, which offers Sunday visits, and the Griffith Observatory, which has astronomical viewing day and night with a heliostat and a 12-inch Zeiss refractor. Examination of the addresses of telescope manufacturers who advertise in *Sky and Telescope* shows that many of them are located in the Los Angeles area; one might wish to inquire about visiting their showrooms or plants, though the one such visit I made was rather disappointing. Any electronics hobbyist will find himself imparadised in the C & H Sales Company at 2174 East Colorado Street in Pasadena; their ads give no hint of the tons of superseded and surplus equipment—Nicad batteries of all sizes, oscilloscopes, scales, bearings, bombsites, computers, and many things whose identity can hardly be determined, plus a certain amount of optical equipment. Although it has nothing whatever to do with astronomy, the visitor to Pasadena will have missed the opportunity of a lifetime if he does not spend a few hours in the Huntington Library's gallery and gardens. Also, anyone returning from Palomar can cut over to the coast and stop at the Mission of San Juan Capistrano, an extraordinarily lovely place even when the swallows are not returning.

Lick Observatory
Mount Hamilton, California 95140

Among the unsolved mysteries of the universe is the impulse that led James Lick, maker and seller of pianos, to spell out in the will in which he disposed of some of his $3,000,000 the terms for the creation of an astronomical observatory. He was not astronomically inclined himself, and had not been the target of one of the great persuaders, such as George Ellery Hale. More than giving the money, though, Mr. Lick picked out the precise site on top of Mount Hamilton; by some feat of divine guidance it turned out to be the best location for an observatory ever found, prior to 1875, and would still be among the best were it not for the lights from the Bay Area, which are sometimes mercifully smothered in fog below. The oft-quoted terms of the deed of trust directed, amid much legalese, that the Trustees "expend the sum of $700,000" [read: $5 million in 1975 terms] for land and for "putting upon such land as shall be designated by the party of the first part, a telescope that shall be superior to and more powerful than any yet made. . . ." Mr. Lick is himself buried beneath the telescope.

Another condition for erecting the Observatory was that Santa Clara County build a "first-class road" to the summit. This road, built in 1876, is still about the same class as it was then and like the drive up Mount Wilson provides ample gratification of the death wish.

The first telescope set up on the mountain was a 12-inch refractor, made by Alvan Clark and thought by him to be one of the best of that aperture ever made; this belonged for a while to another telescope maker, Henry Draper, who traded it back in to the Clarks. Visitors may still look through this instrument, which one of history's keenest-eyed astronomers, E. E. Barnard, used to make "exquisite photographs of comets and nebulae," according to D. J. Warner.

Aerial view across Mount Hamilton towards San Francisco. On the central peak is the dome of the 36-inch Lick telescope, second-largest refractor in the world, and to the right, that of the 120-inch reflector, for many years second in its class also. (Lick Observatory Photograph)

In 1888 the great 36-inch Clark refractor was installed, on a Warner and Swasey mounting. In addition to the crown and flint elements of the achromat, there was a 33-inch correcting lens for photographic work. One of the Clark sons, Alvan Graham, brought this lens out to California himself and apparently offended some people by finding fault with everything but the main optics of the telescope. The 36-inch lens is still carried on the original mounting, in a 57-foot riveted tube that is 4 feet in diameter at the center and surrounded by the same assemblage of cranks, gears, rods and chains with which these early refractors were managed. Four years after its installation, in 1892, it enabled Barnard to discover a fifth satellite of Jupiter, the first addition to that planet's family since Galileo looked at it with his parchment tube and a spectacle lens. The 36-inch has since been used extensively in astrographic work, as at Yerkes, and for the determination of binary orbits, as with the 26-inch Naval refractor. It also provided many spectra for W. W. Campbell's pioneering work on the radial velocities of stars; by 1935 the thousands of determinations (of the speed with which the brighter stars are moving towards or away from us) had made it possible to estimate the movement of the solar system within the galaxy.

In 1895 another important instrument was added to the Observatory; again it was one of 36-inch aperture, but this time it was a reflector that had been built in England in 1881 with a mirror made by Sir Howard Grubb and figured by George Calver. The mounting was made by Dr. A. A. Common and was a strange spidery affair that in some ways resembled a rocking chair, the rockers being its declination gears. This telescope, which had been made for Edward Crossley, Esq. and is named for him, represented one of those fortunate coincidences of emerging technology: Dr. Common had learned how to adapt the new silvering processes to large glass mirrors, and photographic plates suitable for astronomy were more readily available. Aluminized and mounted in a closed tube, this mirror continues to supply long-exposure photographs of gaseous nebulae and spiral galaxies, as well as stellar spectra, many of which have revealed the presence of spectroscopic binaries.

(For additional information on the early history of Lick Observatory, see Donald E. Osterbrock's articles on "The California-Wisconsin Axis in American Astronomy" in the issues of *Sky and Telescope* for January and February 1976. Osterbrock, current

Original mounting of the Crossley reflector, as it came from England. (Lick Observatory Photograph)

Bent-cone English-type mounting with which the Crossley mirror is still used for advanced astrophysical research. (Lick Observatory Photograph)

director of Lick, has reexamined many letters and other documents and has assembled a wealth of information about nineteenth-century astronomers and observatories. His investigations explain much about Yerkes, Washburn, Mount Wilson, and other observatories as well as shedding light on the beginnings of Lick Observatory.)

Although the Observatory acquired at different times a variety of smaller telescopes, including several fine Clark refractors, there was no further addition of major equipment until 1941 when the first of a pair of 20-inch diameter astrographic telescopes given by the Carnegie Corporation was put in place. The war delayed delivery of the second, which was critical to the long-planned survey of the sky. The first 20-inch lens (with four elements, making it in effect a giant camera) was corrected to make its sharpest images in blue light, and the second, in yellow light. Each was designed to cover the same six-degree field, the idea being to register the proper motions of as many stars as possible using faint galaxies as reference points. (For more about the problems and principles of astrographic work, see the discussion of the Flagstaff Naval Observatory). This sky-survey work has been under way since 1947.

The Carnegie 20-inch double astrographic telescope. These two wide-field astro-cameras can cover identical areas of the sky in two wavelengths simultaneously. (Lick Observatory Photograph)

The 120-inch reflector. (Lick Observatory Photograph)

Beginning in 1950, the California state legislature appropriated various amounts that add up to $2,800,000 for the construction of a 120-inch reflector, and this went into operation in 1959. A somewhat dated brochure distributed by the Observatory still describes it as the second largest in the world, which it was, until surpassed by 158-inch mirrors at Kitt Peak and Cerro Tololo (Chile), and by the new Russian 6-meter. Still, its aperture is sufficient to make it an essential tool for collecting light from the more distant galaxies and quasars.

Improvements both in auxiliary equipment and in the devices that measure and interpret the data collected by the telescopes are constantly being made. The Lick

Gaertner automatic measuring engine now operates completely under computer control; this both makes it more accurate and takes a load off the operators. The third version of an image-intensifier for use with the coudé focus of the 120-inch is in place, using liquid-nitrogen tanks to provide a flow of cooling gas. By boiling off the nitrogen under constant pressure, it is possible to provide a stable temperature. This system is also used to cool plates taken at the prime focus of the Crossley reflector, to cut down on reciprocity failure of the film. Early in 1974 a new spectrograph went into operation on the 120-inch; with an image-tube scanner, it can be used by remote control. (For a description of somewhat less comfortable ways of taking the spectra of stars, see the paragraphs on the Harvard Observatory in the 1880s!) The 120-inch is gradually being converted to a completely automatic mode of operation; this sort of automation was pioneered in recent years at M.I.T. (see Part II, Massachusetts: Wallace Observatory). Advances in the technology of infrared spectroscopy are also permitting use of the 120-inch for such a purpose. New high- and medium-dispersion instruments are being developed and tested. Numerous programs are under way either at Lick or at national observatories and in space vehicles; these include astrometric studies, as usual; stellar spectroscopy; variable stars; star clusters; interstellar matter; galaxies; QSOs; in fact, almost everything possible.

One other relatively recent addition is a 24-inch Cassegrain which is used entirely for photoelectric measurements of the brightness of stars.

For many years Lick was an unusual institution in that all offices, laboratories, and supporting equipment were right there on top of Mount Hamilton with the telescopes. But in 1966 new headquarters were set up at the University of California at Santa Cruz. Here the theoretical work is carried out and the computers are housed; here the measuring engine, that determines how far stars have moved across the sky survey plates ("proper motions") in a seven-year interval, is installed in an environment-controlled room to keep its mechanical parts stable.

Visiting: Lick does more to accommodate public interest than any other major observatory; its only rival is Kitt Peak. From June 15 to September 15 the Observatory operates a Friday-evening visitors' program which permits the public to observe with both 12-inch and 36-inch refractors, and at times with other instruments. (The 120-inch has never been used for anything but research). Free tickets, up to a maximum of six tickets per party, may be obtained by writing: Visitors' Program, Lick Observatory, Mount Hamilton, California 95140. Specify first and second choices for dates, and enclose a stamped, self-addressed envelope.

Every day except during university holidays the gallery of the 120-inch reflector is open from 1:15 to 5 p.m., and at the same time in the main building visitors may see the 36-inch refractor and some displays of photographs.

Special tours by groups of astronomy students or amateur clubs may be arranged.

To reach the Observatory, take U.S. Highway 101 through San Jose and turn east at Alum Rock Boulevard. Turn right when you reach S.R. 130 (Mount Hamilton Road).

Accommodations: The bayshore municipalities between San Jose and San Francisco are linked by California Highway 82, which is a sort of strip with pockets of motels and restaurants. Some of these may advertise quite cheap rates, depending on how business is. One might get onto 82 a few miles on the south side of San Jose and head north looking. In Menlo Park and Palo Alto, where 82 is called El Camino Real, there are a number of motels. To the south of Mount Hamilton, east of U.S. 101, the Henry Coe and Humboldt Redwoods State Parks offer camp sites.

Nearby: Of special astronomical interest is the Chabot Observatory (q.v.), and the Lick Observatory recommends the program at the Morrison Planetarium in Golden Gate Park, San Francisco, as well as those at the College of San Mateo and Foothill

The second-largest refracting telescope in the world, made by the Clarks. Modern instrumentation does not conceal its nineteenth-century charm; the dome is beautifully paneled, and has a hardwood rising floor. (Lick Observatory Photograph)

College (in Los Altos Hills). The San Francisco Sidewalk Astronomers are a very active group that makes frequent field trips and offers viewing opportunities to the public; they own an enormous alt-azimuth mounted 20-inch reflector fitted with a wide-field eyepiece for visual observation of galaxies, nebulae, and clusters. San Francisco State University has a very active visitors' program, with a large selection of telescopes for visual use. Ricard Observatory of the University of Santa Clara has a large telescope with a colorful career that is often available for public viewing.

(See entries in Part II under California.)

PART II
A CATALOG OF
U.S. OBSERVATORIES AND
SOME IMPORTANT MUSEUMS
AND PLANETARIUMS

Entries are arranged alphabetically by state, and by cities within states. Figures quoted—whether telescope aperture, elevation above sea level, times, fees, or dates—are mostly those supplied by individual institutions. Equipment will be added or retired, and economic conditions and schedules will change, though no faster than astronomical theory makes textbooks obsolete! Few observatories charge fees; those that do are noted. I have provided information about times of access for all but a few optical observatories; I asked only a few radio observatories about this, since there is no possibility at all of "looking through" a radio telescope unless you are a researcher. Anyone who wants to visit should certainly inquire of the installation he is interested in. The purpose of this catalog is to include every institution of importance to astronomy in the country, whether on account of research, teaching, or public service. Such a catalog must depend on information supplied by such institutions, and these entries may reflect individual eagerness, indifference, or even aversion to publicity. I did my best to even out the coverage, though, and drew on other sources (mentioned in the bibliography) for thousands of additional facts. I welcome all suggestions for revision and all new information. Here are two items of interest to newer observatories: the Nautical Almanac Office of the U.S. Naval Observatory is compiling a new list of observatories; institutions that believe themselves qualified to be listed should send in precise geographical coordinates. Second, beginning and amateur astronomers should be reminded that any person with access to even a small telescope can carry out valuable research by participating in the programs of the American Association of Variable Star Observers, 187 Concord Ave., Cambridge, Mass. 02138.

Alabama

UNIVERSITY OF ALABAMA OBSERVATORY. Dept. of Physics and Astronomy, P.O. Box 1921, University, Ala. 35486. On University Blvd. on the campus. Visiting by appointment only. The University owns a 10-inch J. W. Fecker refractor and several smaller instruments on temporary mounts. These are used primarily for instruction in astronomy, and no research efforts are currently under way.

Other observational telescopes:
Auburn University
NASA-Huntsville
University of Southern Alabama
Wallace Junior College

Larger planetariums:
Robert R. Meyer P., Birmingham Southern College, Birmingham
Florence State University
Rocket City Astronomical Association, Huntsville
W. A. Gayle P., Montgomery

Alaska

CHENA VALLEY RADIO FACILITY. Geophysical Institute, University of Alaska, College, Alaska 99735. Located at Chena Valley. The 18.6-meter steerable dish and the two 8.53-meter dishes set up as an interferometer are currently not being used.

Other observational telescopes:
Alaska Methodist University
University of Alaska

Larger planetariums:
Marie Drake Junior High School, Juneau

Arizona

PHOENIX COLLEGE. 1202 W. Thomas Rd., Phoenix, Ariz. 85073. Just off 11th Ave. and Thomas Rd. The 14-inch Celestron is available at a monthly astronomy open house; call Switchboard Operator (602) 264-2492 for dates and times.

CENTER FOR METEORITE STUDIES. Arizona State University, Tempe, Ariz. In 1960, the National Science Foundation and the ASU alumni association purchased the meteorite collection of the American Meteorite Museum (which was at one time in Sedona, Arizona). This collection contained much material from the Barringer Crater and elsewhere.

UNIVERSITY OF ARIZONA. Tucson, Ariz. 85721. **STEWARD OBSERVATORY,** on the downtown campus, has been holding public evenings for 52 years. No ticket or reservation needed; write for a schedule. Mon near the first and third quarters of the moon except during academic holidays. A lecture by a staff or visiting astronomer at 7:30 p.m. followed by observation with the 21-inch reflector if weather permits, and also with the James refractor. Constellation and star identification from the Observatory roof. Children admitted with adults. Groups of children, sixth-grade and older, accommodated by appointment only for Friday-night viewing through an instructional telescope; enough adults to control them must come along. Since the opening of Kitt Peak, the larger Steward instruments (36- and 20-inch reflectors) have been moved up there, and to them has been added a 90-inch Ritchey-Chrétien reflector. These are not open for visits, though the location may be visited at Kitt Peak (q.v.). All manner of research programs are carried out by the Steward staff, using auxiliary equipment that includes cameras, photometers, coudé spectrograph, spectrograph scanners, image scanners, and so on. Also opening soon on the campus will be the **FLANDREAU PLANETAR- IUM,** the most advanced public educational facility for astronomy in the world, at its completion. Almost 9,000 stars—every single star visible to the keenest unassisted eye from this planet—will be projected on the 50-foot dome, which curves down nearly to eye level to make a realistic horizon. A special 180-degree fish-eye projector will fill the entire surface of the dome with a sharp projected image for special effects. Two hundred other projectors will be arranged around the periphery for other effects. "Omni- phonic" sound effects can be located at any point on the surface of the dome by choos- ing a combination of speakers in that direction. An "evening-sky" planetarium can be entered any time to get an idea of the arrangement of constellations to be expected on that date. Ample space will be available for astronomical exhibits. On the roof there will be a public observatory with its own 16-inch reflector, which will be open every clear night. A bookstore will sell slides, souvenirs, and instruments, as well as astronomical publications. As soon as construction is complete and the bugs have been worked out of this operation, it ought to make one of the most fascinating visits in the world, and unsurpassable when coupled with a day-long side trip to Kitt Peak, about 60 miles SW. Also located on the campus are the offices of the Lunar and Planetary Laboratory (LPL) (see Catalina Mountains Observatory Complex for a description of their instruments). Much of the work for all observatories in the area is carried out in the Optical Sciences Center of the University; this Center is currently involved with the Multiple Mirror Telescope (see Mount Hopkins Station). (See Kitt Peak in Part I for an account of accommodations in Tucson.)

MOUNT HOPKINS STATION. c/o The Smithsonian Astrophysical Observatory, Cambridge, Mass. 02138. Located about 30 miles due south of Tucson, Ariz. Not open for public visits. Equipment currently in place includes a 60-inch reflector that can be used as an f/10 Cassegrain or an f/24 coudé, with a Fourier transform spectrometer and a spectrum scanner. This has been used to look for deuterium emissions in the Orion nebula. There is a 20-inch Cassegrain with a laser for satellite tracking, and one of the SAO's f/1 super-Schmidt Baker-Nunn cameras of 20-inch aperture is also located here. An unusual piece of equipment already used is a 10-meter optical reflector consisting of 248 small hexagonal reflectors, each of 7.3 mm focal length; this is used to search for Cerenkov light produced by gamma rays entering the atmosphere. In collaboration with the Steward Observatory and the Optical Sciences Center of the University of Arizona, the SAO is completing a Multiple Mirror Telescope. Six 72-inch mirrors will be clustered around a core structure, reflecting light to a common focus; aperture should be equivalent to about 175 inches. It will be interesting to see how well this will work; if it is success-

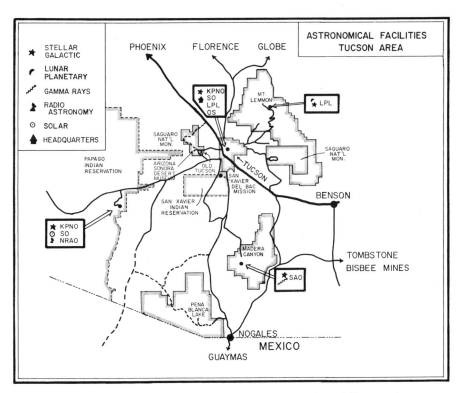

Astronomical Facilities in the Tucson area. (Photo: courtesy Steward Observatory)

Artist's conception of the Mount Hopkins multiple mirror telescope housing. The entire structure will rotate with the telescope, like a huge anti-aircraft turret. (Photo: Smithsonian Astrophysical Observatory)

ful, the 20-foot-diameter area covered by the mirrors might permit resolution comparable to the Russian 6-meter single-mirror telescope.

CATALINA MOUNTAINS OBSERVATORY COMPLEX. c/o University of Arizona, Tucson, Ariz. 85721. This includes **MOUNT LEMMON OBSERVATORY, CATALINA OBSERVATORY,** and **TUMAMOC HILL OBSERVATORY.** These are located about 20 miles northeast of Tucson, and are not normally open for public visits. **MOUNT LEMMON** is 9,200 feet in elevation, and on this site there are telescopes operated by several institutions. A cooperative program of the Universities of Minnesota and California at San Diego operates a 60-inch Astro-Mechanics Cassegrain. This telescope's original mirror (1970) was of aluminum alloy, but the image quality became worse with time; in 1974 it was replaced with a Cervit mirror, whose back was "sculptured" parallel to the parabolic surface to reduce weight. Photometric work and spectroscopy, chiefly in the infrared, is carried out with this. Another identical 60-inch Cassegrain sponsored by NASA was installed on the mountain; its metal mirror will also be replaced. The University of Arizona's Lunar and Planetary Laboratory (LPL) has a 28-inch f/16 Cassegrain on this site, plus another 40-inch Cassegrain. Auxiliary equipment for the two 60-inch reflectors consists of narrow-band photometers, a Fourier transform infrared spectrograph, and other photometers and polarimeters. Nine miles from Mount Lemmon at 8,300 feet is the **CATALINA SITE,** where the LPL's 61-inch Cassegrain/coudé and its 16-inch Schmidt are located. Auxiliary equipment for the 61-inch includes high dispersion spectrographs, planetary and cometary cameras, and a double star micrometer. High-resolution photographs of Jupiter and Saturn have been

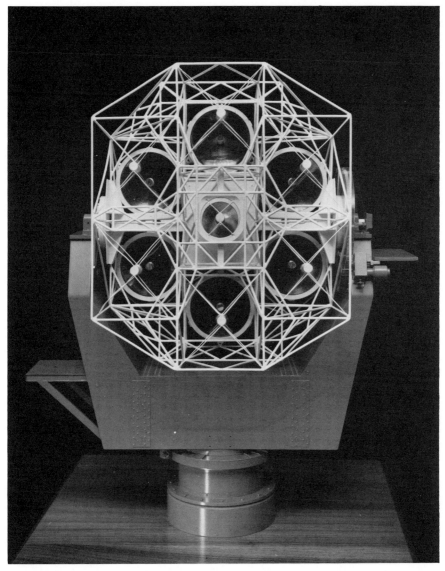

Model of the MMT (Multiple mirror telescope). Each of the six mirrors will be 6 feet across. (Photo: Smithsonian Astrophysical Observatory)

made with this reflector, and their atmospheres have been studied. The Schmidt camera provided photometric and spectrographic studies of comets Kohoutek and Bradfield. **TUMAMOC HILL** has a 16-inch reflector, also operated by LPL. The offices of the LPL are located on the University of Arizona campus, and, under the direction of Gerard P. Kuiper, it has figured importantly in the early stages of space exploration.

BARRINGER METEOR CRATER. Meteor Crater Enterprises, Inc. P.O. Box AC, Winslow, Ariz. Located 19 miles west of Winslow on I-40, 6 miles south of the turnoff. Open 8 a.m. to sundown. In February, 1975, it was unexpectedly closed for unexplained reasons. Ask locally. The meteor crater is worth going a considerable distance to see, and worth paying the admission fees: adult $1.50, ages 13–18 $1, ages 6-12 $.50. First recorded discovery of the crater by white men was in 1871; it is not surprising, in view of abundant evidence of volcanic activity to the west and north, that it was mistaken for a volcano. In 1903 Daniel Moreau Barringer, a mining engineer from Philadelphia, bought the crater and arranged for various scientific investigations that proved it to be caused by a meteor. Guesses at the age of the crater range from 10,000 to 50,000 years. It is 570 feet deep and more than 4,000 feet across. Viewing is from walkways around the museum, which has extensive displays about the crater and about meteorites. There is a gift shop and a refreshment stand.

Other observational telescopes:
Arizona Western College
Glendale Community College
Mesa Community College
Pima Community College
Scottsdale Community College

Arkansas

UNIVERSITY OF CENTRAL ARKANSAS OBSERVATORY. Conway, Ark. 72032. Atop the Science Bldg. Open Tue and Thur evenings and by request at other times. The 16-inch Starliner reflector has camera attachments, a television camera, a spectrometer, and other instrumentation. It is a Newtonian/Cassegrain. The telescope is used for photometry, but is mainly for use by students, to whom it is available at all times other than public hours. There is a physics degree with emphasis on astronomy.

UNIVERSITY OF ARKANSAS. Physics Dept., Fayetteville, Ark. 72701. There will be a small observatory here, with a 16-inch reflector, which will be completed during 1976–77. There will probably be public hours, and some research in photoelectric photometry.

Other observational telescopes:
Henderson State College
Larger planetariums:
University of Arkansas

California

BIG BEAR SOLAR OBSERVATORY. California Institute of Technology, Pasadena, Calif. 91125. Located on North Shore Drive, Big Bear Lake. Visits by arrangement; public visiting hours every Fri afternoon. This unusual observatory is located on a tower whose foundations are an island in Big Bear Lake; the water level of this irrigation reservoir can vary drastically, and at times the island is submerged. The purpose of the location is to take advantage of the stable air-layering that occurs over the surface of a cool body of water during the day; this eliminates much turbulence and makes possible solar observations superior to those at a site such as Mount Wilson. Furthermore, the 6,700-foot elevation and the absence of smog provide excellent transparency. Three independently guided vacuum telescopes are mounted on a massive stubby fork, which need be capable only of movement within 24 degrees of declination. One 6-inch vacuum refractor provides full-disk hydrogen-alpha patrol; a 10-inch vacuum refractor feeds a triple camera for simultaneous 35 mm films in various wavelengths, such as hydrogen-alpha, white light, and K line. An f/50 Gregorian reflector with a 65 cm mirror and an evacuated tube reflects light through the coudé system to a spectrograph, and also feeds two filters plus 35 mm movie cameras. A 6-inch piggy-back refractor gives full-disk white light. Auxiliary instruments include magnetograph, spec-

Big Bear Solar Observatory, usually reached by boat, holds various solar telescopes. Cool water in the reservoir keeps the air steady during the day. (Photo: courtesy California Institute of Technology)

trograph, and universal birefringent filter. All manner of solar studies can be carried out. The San Bernardino Mountains contain numerous campsites with fine dark skies.

LEUSCHNER OBSERVATORY. Astronomy Dept., University of California at Berkeley, Calif. 94720. Located north of Lafayette, Calif., 25 minutes drive east of Berkeley. No scheduled public hours, although special sessions can be arranged for educational groups through the director. There are two reflectors, a 20-inch Cassegrain and a 30-inch Ritchey-Chrétien instrumented for direct photography and for photoelectric photometry (UBV RI and uvby H-B). The 30-inch also has a two-channel sequential spectrum scanner controlled by a small computer as well as a more conventional spectrograph. The Observatory is used primarily as a graduate student research center as well as for teaching purposes. It is also used to test more complicated instrumentation for use on larger telescopes. The Observatory was founded in the 1890s and was originally known as the Students' Observatory. It was renamed Leuschner Observatory in 1954 and was moved off campus to its present location in 1965. There is still a 10-inch reflector and a 4-inch astrograph on campus located on top of Campbell Hall for undergraduate teaching purposes. There is also a 5-inch refractor. Scientific scheduling requests on the large instruments off campus are handled in two-week periods, about three weeks in advance.

OWENS VALLEY RADIO OBSERVATORY. California Institute of Technology, Pasadena, Calif. 93513. Located at Big Pine, Calif. No regular visiting, but special arrangements can be made in advance with the director by recognized or qualified groups. Equipment consists of two 90-foot dishes and one 130-foot, used singly and in interferometer combinations between wavelengths of 1 m and 1 cm. A 10-meter telescope for wavelengths to 1 mm is under construction. Research in Very Long Baseline Interferometry is carried out very actively here, with a good deal of effort expended in developing the instrumentation for this (see Green Bank in Part I and Haystack). Observations at 40-meter wavelength were carried out when Comet Kohoutek's tail occulted a bright radio source, but CO absorption was not observed. In galactic radio astronomy, there have been projects involving observation of the neutral hydrogen line, OH, and formaldehyde. Extragalactic work uses the VLBI techniques, and there are both continuum and spectrographic radio observations of extragalactic objects. This Observatory puts special emphasis on the training of graduate students. Touring amateurs may find campsites in the Bishop, Calif., area.

FRANK P. BRACKET OBSERVATORY. Dept. of Physics and Astronomy, Pomona College, Claremont, Calif. 91711. The college also owns a planetarium, in the Millikan Laboratory at 6th and College Ave., which offers free lectures and shows at 8 p.m. on the first academic Mon of each academic month. A 6-inch refractor is available for viewing after the shows. Special shows and viewing can be arranged if the staff is available; call R. J. Chambers (714) 626-8511, ext. 2945. The planetarium uses a Viewlex "Mercury" in a 20-foot dome seating forty-eight. The Observatory also owns a large variety of other instruments: a 22-inch Celestron, a 40-foot horizontal solar telescope, a 12-inch Porter-designed reflector, an Askania 3-inch bent transit, a Wild T-2 theodolite, and a satellite receiver. Lunar occultations, photometry, photography, and some determination of minor planet positions are among the observations that are made here. But the principal purpose of the program is education. Physics majors can concentrate in astronomy, and many go on to graduate schools; 60 percent of the student body takes an astronomy course. According to Professor Chambers, "Contact with active local amateur groups is encouraged." In addition to the telescopes, there is an astrophysics laboratory with measuring equipment for studying photographic plates and an extensive astronomy library. These are located in the Millikan Laboratory. Sites in the San Gabriel or San Bernardino Mountains provide better viewing for amateurs than anywhere in the Los Angeles basin (see Mount Wilson in Part I and Table Mountain).

CLARK LAKE RADIO OBSERVATORY. University of Maryland, P.O. Box 73, Borrego Springs, Calif. 92004. To judge from the extent of the visitors' program at the Maryland optical facilities, visits can probably be arranged. The description of the radio telescopes, supplied by Maryland, is so well written and so informative about radio astronomy in general that, rather than adhere to the general format of these entries, correlating other available sources of information, it will be quoted here verbatim: "Located on a dry lake bed, CLRO is dedicated to the reception and investigation of radio signals reaching earth from other parts of the Solar System and from deep space. Unlike most other radio astronomy observatories, CLRO has no 'big dish' antennas. At the relatively low radio frequencies received at this installation, a parabolic big dish would have to be many times the size of the world's largest—the 1000′ diameter dish at Arecibo, Puerto Rico. CLRO's antennas, therefore, are of the fixed array type. The receiving antennas are formed of a number of elements, carefully spaced over the surface of the dry lake. The elements are linked together with transmission lines in such a way that the small amount of radio energy picked up by each element will be added to that received by all the other elements to produce a signal large enough to be detected by the receiving equipment housed in the central building. There have been several generations of arrays at Clark Lake. The original antenna was simply a number of bare wire dipole elements hung between wooden posts, rather resembling a big fence. The next receiving system is composed of sixteen large white log periodic elements evenly spaced over a two-mile east-west base line (this antenna is owned and operated jointly by the University of Maryland and the NASA-Goddard Space Flight Center); because of its base line length, the end elements are built six inches higher than the center elements in order to compensate for the curvature of the earth's surface. Recently completed is a much bigger array of 720 conical spiral elements in the form of a huge 'T'; this array looks like a forest of 25-foot high wire Christmas trees. A large antenna has been built and is operated by the National Oceanic and Atmospheric Administration and the University of Iowa. The University of California at San Diego also has a radio telescope in operation at CLRO; this consists of 256 TV antennas pointing out into space and relaying their findings to UCSD over a telephone line. While all of the elements of each of these arrays are in fixed positions on the surface of the dry lake, their received radio energy is combined to form a resultant 'beam' pattern that can be electronically steered to point at different positions in the sky. The steering of the new conical spiral array, as well as the UCSD array and the NOAA-Iowa array, is under computer control and it is possible to point each of them to a different part of the sky every few thousandths of a second. Transmission lines bring the received signals from the antennas into the observatory building in the center of the dry lake. Here a number of different types of receivers with display and recording systems are in operation. Narrow bandwidth receivers are used to observe signals in the quiet windows which fortunately still exist between the areas of man-made interference. Wide band receivers are used to scan a range of radio frequencies which extends from below the Citizen's Band up into the ranges used by TV. (Since radio signals are never transmitted from CLRO there is no possibility of the operations here interfering with radio or TV.) A recorder is used to make a display of all the signals received in this wide range of frequencies. In addition to the radio emissions from space, signals from Citizens' Band transmitters, mobile communications systems, and TV stations also are received. Variations in the earth's ionosphere cause many of these signals to change randomly, and only careful analysis by trained observers can distinguish between this man-made interference and significant signals from space. For this reason, even the most powerful computers can't always be programmed to analyze radio astronomical data; experienced observers have to analyze the records inch by inch to determine which signals are of astronomical origin. So far, the antennas have picked up radio signals from the sun and the solar corona, from the planet Jupiter, and from several hundred radio stars. (The antennas are also able to pick up radio noise caused by automobile ignition systems, so the arrival of visitors to the site is usually noted in advance. Also, the fence-like array spread over the dry lake surface has collected several motorcycles, cars, and at least one dune buggy; in every case of such physical conflict to date, the antennas have won.) Data from CLRO has helped increase man's understanding of the structure of the sun's corona—that halo of gases surrounding the sun which can be seen by the naked eye only during total eclipses. Continuing studies of radio emitting stars, such as the one in the center of the Crab Nebula, provide clues to the mechanics of the birth, life, and death of stars. (This particular star exploded in the year 1054 A.D., forming a supernova so bright it could be seen in the daytime, as was recorded by Chinese astronomers of that time.) A constant watch was kept at CLRO during the flight of Apollo 12 to insure that radiation from bursts of energy from the sun didn't endanger the astronauts. Continuing observation of solar activity also increases our knowledge of the processes in the sun that give rise to sunspots and related activity; these observations are conducted 365 days a year, rain or shine, as the use of radio equipment enables us to 'see' through clouds."

(Radio telescopes similar to those used at Clark Lake are illustrated with the entry on the University of Texas Radio Astronomy Observatory.)

UNIVERSITY OF CALIFORNIA AT DAVIS OBSERVATORY. College of Letters and Science, Davis, Calif. 95616. A 12½-inch Cave reflector is mounted in a dome on top of Hutchison Hall; an 8-inch Springfield-mount reflector and a 4-inch refractor are on the roof of the Physics Bldg; the last of these is an 1875 Alvan Clark product. Thurs nights the Physics Bldg roof is open, with slide shows and minilectures as well as observing; on alternate Wed the Hutchison Hall dome has been open to the public. The telescopes are mainly for undergraduate use and for viewing by the UCD Astronomy Club. But sometimes there are projects by physics students, and there is a photometer for use on the 12½-inch to observe eclipsing binaries. There are some astronomy courses within the Physics Dept.

GOLDSTONE TRACKING STATION. Jet Propulsion Laboratory, NASA, 4800 Oak Grove Dr., Pasadena, Calif. 91103. Located about 45 miles north of Barstow, Calif., at Fort Irwin. Individual visitors and group tours welcome at any time, but it is recommended that prospective visitors call the Public Relations Office (714) 288-8555 or, during weekends, the Main Gate (714) 288-8250 in advance. The NASA Deep Space Network maintains several antennas here. The largest is at the Mars complex, which is part of a world-wide system; the antenna here is the 210-foot dish completed in 1966 primarily for the purpose of tracking and communicating with spacecraft. This big paraboloid with a Cassegrain feed has, however, been used for a variety of radio and radar astronomy projects: it has observed pulsars, recorded Jupiter's radio signals, timed occultations of radio sources by the moon, observed spectral lines, and participated in Very Long Baseline Interferometry experiments (see Green Bank in Part I and Haystack). As early as 1958, Goldstone had an 85-foot equatorially mounted paraboloid, designed to track Pioneer space vehicles to the moon; this is now mainly used in VLBI projects. A similar antenna serves for lunar and solar occultation observations. A third 85-foot dish succeeded in 1961 in bouncing radar signals off Venus, providing information on its rotation rate and helping to make the measurement of the Astronomical Unit more precise. Now it observes pulsars and is used for occultation and interferometric work. A smaller 30-foot dish observes Venus and Jupiter. All antennas have a Cassegrain feed arrangement with a focussing secondary. The Mars antenna, in addition, operates receivers between 7.8 and 8.7 GHz with a 25K system temperature. It is, at once, the largest and most sensitive, fully steerable radio telescope in the United States. The secondary reflector can be indexed to select any of the receiver

The 210-foot Mars antenna at Goldstone. Note the three feedcones for frequent switching. (Photo: courtesy NASA/JPL)

feeds. In addition, an arrangement using a dichroic flat permits simultaneous use of 4-cm and 13-cm receivers. While most of the radio science antenna time supports NASA-funded research at JPL, the antenna is made available to qualified astronomers through the Radio Astronomy Experiment Selection Panel.

RADIO ASTRONOMY LABORATORY. University of California, Berkeley, Calif. 94720. Hat Creek, Calif. SR 89. Visits by request. Principal instruments include an 85-foot dish used for radio spectroscopic studies of the Milky Way and a two-antenna interferometer, each 20 feet in diameter, with a maximum baseline of 1,000 feet. The latter pair is used for high spatial resolution studies at wavelengths from 1.5 cm to 1.5 mm. This center is working on most of the frontiers of radio astronomy, and studies in progress include spectroscopy of interstellar atoms and molecules; observations of planets; quasars; and astrometry. Interstellar OH emission was first observed here with the 85-foot telescope, and interstellar water and ammonia were discovered with the 20-foot dishes. Nearby Lassen National Park offers excellent seeing conditions for vacationing amateurs.

IRVINE OBSERVATORY. University of California at Irvine, Calif. 92664. The University owns 8-inch and 5-inch Celestrons, which are used for the undergraduate course on observational astronomy. No scheduled public hours. There are some lecture courses given in astrophysics.

FOOTHILL COLLEGE OBSERVATORY. 12345 El Monte Rd., Los Altos Hills, Calif. 94022. Open every night during the summer and Fri nights throughout the year, as well as for particular astronomical events. Telescopes for visual observing include a 16½-inch reflector, a 6-inch refractor, a 4-inch solar prominence telescope, a 5-inch richest field telescope, and various finders. There are also astrocameras. Some photometry of variable stars is carried out. This is the largest publicly owned telescope in the Bay Area, and is operated by the Community Services of Foothill College. Operations are coordinated with those of a planetarium (also at Foothill) and other educational programs. The solar prominence telescope here was the first owned by a public institution.

GRIFFITH OBSERVATORY. 2800 East Observatory Rd., Los Angeles, Calif. 90027. Take Vermont Ave. north into Griffith Park. Open free during the school year: Tue–Fri 2–10 p.m.; Sat 10:30 a.m. to 10 p.m.; Sun 1–10. Summer: Mon–Fri 2–10 and same Sat-Sun schedule as above. Planetarium shows every day: $1.50 adult, $1 youth, $.50 children. The Observatory with its 12-inch Zeiss refractor is open to the public every clear night; the triple coelostat shows the sun's surface when the sun is out and in good viewing position. Numerous exhibits on the nature of light, radio astronomy, the Tesla coil, etc., and a weather satellite receiving station. There are various items of considerable historical interest, including small antique instruments and the core removed from the 200-inch Hale Telescope mirror. Bookstore and shop. This institution, owned by the City of Los Angeles, specializes in public education and scientific entertainments. The staff carries out studies of galaxies, archaeoastronomy, stellar evolution, galactic structure, meteoritics, and flare stars. This is one of the best centers for popular and amateur astronomy in the country. For amateur observing sites outside the city, and campgrounds, see descriptions of the Hale Observatories in Part I and Table Mountain.

MILLIMETER-WAVE OBSERVATORY. The Aerospace Corporation, Electronics Research Lab, P.O. Box 92957, Los Angeles, Calif. 90009. Located at Los Angeles Air Force Station in El Segundo. The 4.57-meter paraboloid is used for continuum and spectral-line radio astronomy at short millimeter wavelengths. Studies that have been or will be carried out with this antenna include galactic CO emission; CO and HCN emission associated with H II regions; CO observations of interstellar dust clouds; CO emission of supernova remnants; and searches for new interstellar molecules.

UNIVERSITY OF CALIFORNIA AT LOS ANGELES. Dept. of Astronomy, UCLA, Los Angeles, Calif. 90024. The high-powered astronomy program here offers numerous courses for undergraduates and graduate students; prospective visitors should get in touch with the Dept. of Astronomy. The following description of research facilities is taken from their brochure: "A small student observatory on campus is equipped with meridian transits and with visual photographic telescopes that have apertures ranging up to 16 inches. The instruments and their accessories are used primarily for instructional purposes. At an off-campus site, the Department has installed a 24-inch Cassegrain reflector equipped for photographic, photometric, spectrographic, and spectrophotometric work. Optical equipment of the Lick Observatory and radio equipment of the Hat Creek Radio Observatory are used by qualified advanced graduate students. We also have access to the observing facilities of the National Radio Astronomy Observatory and of the Kitt Peak National Observatory. Modern equipment for the reduction of observations made with these and other instruments is available on campus. The excellent computing facilities of UCLA, including an IBM 360/75 computer, are available to staff and students. All of this equipment provides an unusually fine opportunity for spectrographic analysis and for theoretical calculations. Through the cooperation of closely related departments, opportunities are available for research in geophysics and planetary physics and for the design and construction of experimental equipment for space vehicles."

CHABOT OBSERVATORY AND PLANETARIUM. 4917 Mountain Blvd., Oakland, Calif. 94619. Write for announcements and maps. Coming from north or south on US 580-50, take Seminary Exit and look for Planetarium signs. Also, just off SR 13, coming from Berkeley. From Berkeley and downtown Oakland, Bus 15D goes past the Observatory; and from High St., take Bus 79. The Chabot Observatory has long been a center for amateur meetings and for popular astronomical education. The several telescopes, which include an 8-inch Clark refractor and a 20-inch refractor made by John Brashear, the third largest he ever made, are all available for public use. Open every Fri and Sat night at 7:30. Planetarium admissions: adults $.50, children $.25—includes film, planetarium program, demonstrations, and observation through a large telescope. Extra programs scheduled for groups of twenty-five or more. Special exhibits and displays. The Observatory sponsors a Telescope-Makers Workshop offering instruction and assistance in making reflecting telescopes. The resources of this very active center are so extensive, varied, and changeable that one should probably visit regularly in person to keep up with what is going on.

JET PROPULSION LAB. California Institute of Technology, 4800 Oak Grove, Pasadena, Calif. 91103. Public visits last Sun of each month, 1–4 p.m. Exhibits of full-sized planetary and lunar unmanned space craft, including examples of the first such vehicles ever launched. This is the NASA facility that has pioneered in space technology since 1944 and in radar astronomy (at the Goldstone Deep Space Communications Complex). Its satellite fly-by program is scheduled for years on into the future.

TABLE MOUNTAIN OBSERVATORY. Jet Propulsion Laboratory, 4800 Oak Grove Drive, Pasadena, Calif. 91103. Near Big Pines Ranger Station, Angeles Crest Highway (SR 2), Wrightwood, Calif. 92397. Group visits arranged; call (213) 354-5708. A 24-inch Cassegrain/coudé is used for photoelectric photometry, low-dispersion spectrophotometry, and high-dispersion planetary spectroscopy. A 16-inch f/50 Cassegrain is used for direct photography. An 18-foot and a 10-foot millimeter-wave radio telescope form a two-element interferometer. The emphasis in these observations is almost entirely on the solar system, with photometric and spectroscopic studies of both airless bodies and of planetary atmospheres, and radio interferometric observations of Venus,

Jupiter, and Saturn. A description especially provided by the Observatory states: "The observatory is located on the heavily forested 7,500-foot summit of Table Mountain in the northeastern part of the San Gabriel range, near the town of Wrightwood, California. Greater Los Angeles is about 40 airline miles to the west-southwest. The observatory site lies within the boundaries of the Angeles National Forest, and is occupied under permit from the U.S. Forest Service. For many years, Table Mountain was the site of one of the stations set up by the famous solar astronomer C. G. Abbot, of the Smithsonian Institution, to measure the solar constant. It was considered as a potential location for the 200-inch telescope, but was rejected because the San Andreas fault runs within a mile of the mountain top! Scattered light from Los Angeles is well screened by the bulk of the San Gabriel mountains. Although the towns of Victorville, Wrightwood, and San Bernardino can be seen on direct line-of-sight from the observatory, skies are generally dark, so useful photometric observations can be made in the visual and near-infrared wavelengths. A popular ski area is directly adjacent to the observatory. As might be surmised, heavy snowfalls occur frequently between December and April but rarely force observatory operations to halt. Precipitation in any form is extremely rare between June and October. Winter temperatures are generally around 20 degrees F. and rarely go below 0 degrees F. Summer daytime temperatures average in the eighties, and night-time temperatures in the sixties." Los Angeles area amateur groups frequently use the large parking lot at the nearby Table Mountain Ski Lodge for star parties. A campground maintained by the U.S. Forest Service is located adjacent to the ski area.

LOCKHEED PALO ALTO RESEARCH LABORATORY. Lockheed Missiles and Space Company, Palo Alto, Calif. 94304. Groups from this laboratory participate in projects in space astronomy and in atmospheric physics, and scientists there have built a spar telescope for solar observing. As originally constructed, a 5-inch steel tube had radial struts welded to it, and this in turn was enclosed in a 28-inch tube. The idea was to provide an enclosed support that could carry and protect at least four instruments for simultaneous solar observations. Flare patrol and studies of the sun with very narrow bandpass filters have been carried out; recently there have been high-resolution filter-graphic studies in hydrogen-alpha light, and a multislit spectrograph has permitted hydrogen-alpha line study during flares. The laboratory now owns a video magnetograph and a computer-controlled microdensitometer.

LA POSTA SPACE GEOPHYSICS RESEARCH FACILITY. Naval Electronics Laboratory Center, San Diego, Calif. 92152. A Cassegrain-feed 18.3-meter paraboloid observes the continuum of the solar radio spectrum, as do two smaller dishes.

SAN FERNANDO OBSERVATORY. In 1976, the Aerospace Corporation gave this facility to California State University at Northridge. Located at Van Norman Lakes in the San Fernando Valley. Equipment includes a 24-inch evacuated telescope and spectroheliograph, a 12-inch spar telescope, several 6-inch refractors, and two radio telescopes for 3 and 10 cm observations. Observations specialize in solar activity and dynamics, particularly magnetic field and chromosphere phenomena. The first time-lapse studies of solar magnetic fields were made here. There is a videomagnetograph that obtains real-time magnetic data for storage or display. Routine visual observations and whole-disk photography are carried out in white light and hydrogen-alpha light. Although the 24-inch is one of the largest instruments anywhere devoted primarily to solar research, astronomers associated with this Observatory also run programs at Kitt Peak, with the McMath telescope, and participate in experiments and observations in space astronomy. The location of the Observatory is interesting, being on a peninsula that extends from the north shore of Upper Van Norman Lake; as at **Big Bear Lake** (q.v.), the water helps stabilize seeing conditions during the day, since the air currents that occur over sun-warmed ground do not develop. The 24-inch reflector is in an odd-shaped dome that looks like a moon-lander, with petals that fold back from above. A 6-inch hydrogen-alpha refractor occupies another tower, whose cylindrical upper portion is lowered during observations.

MORRISON PLANETARIUM. California Academy of Sciences, Golden Gate Park, San Francisco, Calif. 94118. Programs mid-June through Labor Day, daily at 12:30, 2, 3:30, and Wed and Thur at 8 p.m. Rest of the year, daily at 2 plus Wed and Thur at 8 p.m. Additional afternoon shows weekends and holidays. Admission: adult $1.50; under 18, $.50. This is in addition to admission fee to the museum; no museum fee for night shows. Building open 10–5, later in summer; children and those over 64 enter free; 12–16 yrs. $.25, adult $.50. Numerous exhibits on the natural sciences, including a Foucault Pendulum. The Planetarium seats 375 and is extremely well equipped.

SAN FRANCISCO STATE UNIVERSITY OBSERVATORY. Dept. of Astronomy and Physics, San Francisco, Calif. 94132. In a sliding-roof structure on top of the Physical Science Bldg. Open for public observation Tue, Wed, Thur evenings. Instrumentation consists of a 16-inch Optical Craftsman Newtonian-Cassegrain, a 10-inch Celestron, a 10-inch Cave Cassegrain, an 8-inch Celestron, a 5-inch Schmidt Camera, and a 4-inch Unitron refractor. These are primarily used for student research projects; the University has a very active astronomy program, with a BA and MA in astronomy and also in planetarium training; few places offer the latter specialty. Photoelectric and photographic photometry of variable stars is also carried out. An amateur thinking of buying a telescope would find a visit on a public night worthwhile, since so many manufacturers are represented in the telescope collection. There is also a Spitz Systems 512 Planetarium seating 48; lectures here at noon Wed, and by appointment. As part of its instructional program, the Department participates in the Planetarium Institute, a group of six Bay Area Planetariums.

THE SAN FRANCISCO SIDEWALK ASTRONOMERS and THE SAN FRANCISCO AMATEUR ASTRONOMERS. These two groups currently meet at the Josephine D. Randall Junior Museum, 199 Museum Way, San Francisco, Calif. 94114. Phone: (415) 863-1399. The Sidewalk Astronomers own a variety of telescopes with picturesque names, the largest of which is a 20-inch alt-azimuth reflector with a wide-field eyepiece. This instrument is transported by trailer and panel truck to dark-sky sites in the mountains.

RICARD OBSERVATORY. University of Santa Clara, Santa Clara, Calif. 95053. Public open houses Tue and Fri nights, 8:30–11 and Sun afternoon 1–5. The 16-inch objective of the refractor began a picturesque existence in 1882 at the end of a telescope erected in Rochester, New York, at the expense of a patent medicine manufacturer, H. H. Warner, and for the use of Lewis Swift. Swift, a self-taught observer, went on discovering comets until 1899, when he was 79 years old, after which his sight failed. For the whole story, see Joseph Ashbrook's fine account in *Sky and Telescope* for June, 1972. The lens survived transcontinental travel, fire, windstorm, and thousands of hours of public observation, spending almost fifty years atop Echo Mountain (in the San Gabriel mountains north of Los Angeles) serving in part the function of a tourist attraction. In response to a questionnaire, the Observatory at Santa Clara, curiously, lists nothing at all under "Instruments of Historical Interest!" The Observatory owns another Clark telescope, of 8-inch aperture, plus some seismographs and a weather station. Visitors here have the unusual opportunity of frequent access to what was at one time the third largest refractor in the country.

STANFORD SOLAR OBSERVATORY. Stanford, Calif. 94305. This is not open for visits. The facility specializes in solar magnetic fields, and is equipped with the Babcock Solar Magnetograph.

STANFORD RADIO ASTRONOMY INSTITUTE. Stanford, Calif. 94305. Astronomers, student groups, etc., interested in visiting should write to the Institute. Equipment consists of a 32-element cross antenna and a 5-element array of paraboloids with a total collecting area of 600 square meters and a 17 arc-second resolution. These equatorially mounted 60-foot dishes can be used to synthesize various apertures, and are employed in many kinds of radio observations.

STONY RIDGE OBSERVATORY. Stonyridge Amateur Club, 10508 Ormond, Sunland, Calif. 91040. Address may change frequently. Located in the Angeles National Forest. No regular public hours, but field trips for schools and clubs may be arranged. The 30-inch Newtonian/Cassegrain telescope is also used by the University of Southern California and the Pasadena City College for instructional and research purposes. Contact may be made with current club officers by writing c/o these places.

Other observational telescopes:
California State Universities at:
   Dominguez, Fullerton, Humbolt,
   Los Angeles, Sacramento, San
   Bernardino, and Stanislaus
Allen Hancock College
American River Junior College
Canada College
Chaffey College
College of the Redwoods
Citrus Community College
City College of San Francisco
Crafton Hills College
Columbia Junior College
Fullerton Junior College
Glendale Community College
Grossmont Community College
Los Angeles Harbor College
Los Angeles Valley College
Merced College
Monterey Junior College
Moorpark College
Orange Coast College
Palomar College
Pasadena Community College
Pepperdine University
Rancho Santiago Community College
Rio Hondo College

Saddleback Junior College
San Joaquin Delta College
San Jose City College
Santa Barbara City College
University of California at Santa Barbara
San Diego State College
College of San Mateo
Ventura College
Yuba Community College

Larger planetariums:
Millikan P., Chaffey College, Alta Loma
Bakersfield College P., Bakersfield
Southwestern College, Chula Vista
De Anza College, Cupertino
Fremont P., Hopkins Junior High
   School, Fremont
Garden Grove Unified School District
Chabot College, Hayward
Pasadena City College
J. Frederick Ching P., Salinas
San Luis Obispo High School
Palomar College, San Marcos
Tessman P., Santa Ana College,
   Santa Ana
West Valley Junior College, Saratoga
Mt. San Antonio College, Walnut

Colorado

DENVER MUSEUM OF NATURAL HISTORY. City Park, Denver, Colo. 80205. At the corner of Montview Blvd. and Colorado Blvd. The **CHARLES C. GATES PLANETARIUM** has shows Tue–Fri in winter at 1 p.m., Sat and Sun at 2 and 4, and Wed and Sun nights at 8. In summer, the weekday schedule includes shows at 11 a.m., 2, and 4 p.m. Admission: adult $1.25, children $.75. The Museum itself is open most days all day long. It has many natural science exhibits, a Foucault Pendulum, a Space Shop, and a book shop. The Museum also owns a 22-inch Celestron, which is located at Jefferson County Laboratory School on a site 40 miles west of Denver; it is not open to the public at this time.

JOINT INSTITUTE FOR LABORATORY ASTROPHYSICS. National Bureau of Standards and the University of Colorado, Boulder, Colo. 80302. This center is not exactly an observatory, but its programs are carried out frequently using instruments of, or located at, other observatories. For example, one member is developing a Cassegrain spectrograph to be used on the Sommers-Bausch 24-inch, and another has completed a 90-inch equivalent lunar-ranging telescope to be used at Haleakala Observatory (Hawaii). Research programs include stellar structure and pulsation, radiative transfer, non-LTE stellar atmospheres, interstellar matter, X-ray sources, solar physics, and stellar spectroscopy.

CHAMBERLIN OBSERVATORY. University of Denver, Denver, Colo. 80210. In Observatory Park at Warren and Filmore Sts. Open Tue and Sat nights; phone: (303) 753-2070. The 20-inch refractor has a lens figured by Alvan Graham Clark and mounted by Fauth and Co.; it has been in use since 1894, and still tracks the stars by means of its weight-driven clock. The front element of the lens can be reversed for photographic work, following a method invented by the Clarks that obviated a third correcting lens. The Observatory is headquarters for the Denver Astronomical Society, and serves mainly for teaching and public viewing. Nowhere else in the country is a large Clark refractor so accessible to the public; this is its principal use nowadays. The photograpic configuration designed into the reversible lens is never used. (With inexpensive catadioptric telescopes available for photography, it would be foolish to meddle with the optics of this fine instrument.) In nearby Boulder one may visit the Sommers-Bausch Observatory and the Fiske Planetarium; numerous campsites are available to the north and west.

SOMMERS-BAUSCH OBSERVATORY and **FISKE PLANETARIUM.** University of Colorado, Boulder, Colo. 80302. At the intersection of Folsom and Regents Streets. Visitors' nights at the Observatory Fri during the school year, weather permitting. The 24-inch Cassegrain/coudé is used for spectroscopy of early-type stars. The Fiske Planetarium has just opened; inquire about its schedule. At a cost of close to two million dollars, one of the best-equipped planetariums in the world has just been completed adjacent to the Observatory; it will seat 213 people under a 65-foot dome. Among the special effects possible are the projection of colored constellation drawings, and a wide selection of horizon effects. The main projector is a Zeiss VI. The sound system is especially responsive and flexible. On display in the lobby is the old Bausch telescope, a 10½-inch Clark refractor built by George Sagemuller in 1912. This telescope was originally on top of the Bausch and Lomb Building in Rochester. It really ought to be put back into use for observing.

HIGH ALTITUDE OBSERVATORY. National Center for Atmospheric Research, Boulder, Colo. 80302. Stations at Climax and Horse Creek. Two coronographic telescopes of 5- and 16-inch aperture, equipped with spectrographs and magnetographs, continue the program of solar observations at Climax initiated by Harvard College Observatory; this is at the summit of Fremont Pass, 13,300 feet above sea level. In 1970, a 10-inch vacuum telescope was put in service at Horse Creek.

U.S. AIR FORCE ACADEMY PLANETARIUM. Colorado Springs, Colo. Leave I-25 at North Gate Blvd. and follow Blvd., then Academy Rd. to the Cadet Area. Free public shows at 2 and 3 p.m. Sat and Sun during academic year, except home football games Sat. Christmas show daily at 2 and 3 p.m. during Christmas season (Dec 14-31, except 25th). Visits any day from May 24 to Sept 1 at 12, 1, 2 and 3:30, and Jul 7 to Aug 15 at 10:30. Write for a leaflet. Minimum for special arrangements: 40. Capacity is 350. This is the only large planetarium at a military installation. Their leaflet asks that children under four not be brought. Persons traveling north or south might find this a pleasant break on a summer afternoon, with four or five free shows offered daily at a major planetarium.

MOUNT EVANS OBSERVATORY. University of Denver, Colo. 80210. I-70 to Idaho Springs; then Mt. Evans Rd. past Echo Lake to summit. No regular tours, but daytime visits in summer months for groups of about ten can be arranged. The 24-inch Cassegrain has a Ritchey-Chrétien configuration with a "wobbling secondary." The high elevation is favorable for infrared astronomy. "At an altitude of 14,100 feet," writes the respondent to a questionnaire, "Mt. Evans is the highest permanent astronomical observatory in the world. The present telescope, built by Ealing-Beck, Ltd., of Watford, England, was installed in 1973. The public has access to the summit (where there is a commercial guest house and concession stand, unconnected with the Observatory) via paved road from roughly mid-June through Labor Day. At other times the road is closed because of the danger of snow. Prior to the installation of the Observatory, the site was used by the University of Denver's High Altitude Laboratories for cosmic ray and environmental research." The Mt. Evans road is among the very highest in the world accessible by automobile (though in winter and spring one goes by helicopter or on foot). Unacclimated visitors will feel light-headed and should move slowly to avoid palpitations and short breath. Amateurs may find observing sites at Echo Lake and Summit Lake.

RADIO ASTRONOMY OBSERVATORY OF THE UNIVERSITY OF COLORADO. Dept. of Astro-Geophysics, Boulder, Colo. 80302. A steerable spectrographic interferometer and a fixed-array interferometer are the basic instruments at this observatory, which is an NSF-supported research activity dedicated primarily to continuous recording of the long-wave (8–80 MHz) radio emissions from Jupiter and the sun. The facility has been in nearly continuous operation since 1959. Such a long-term monitoring program is essential for Jupiter observations, since Jupiter emits characteristic spectral events which develop through an eleven-year cycle. Other short-term research activities have included a variety of radio interferometers, radio polarimeters, antenna-design tests, and digital techniques for detecting radio pulses from Jupiter and terrestrial spherics. The observing site frequently has served as a base station for Very Long Baseline Interferometry using magnetic tape synchronization techniques. Modern data processing techniques are employed to obtain radio spectra from single-channel observations, to identify and correlate Jupiter pulses recorded at remote sites, to map the apparent source positions of solar radio outbursts, and to catalog solar and Jovian emissions for subsequent statistical and physical analyses. The observing equipment is located on beautiful, mountaintop meadows within view of the snow-capped Continental Divide. Routine radio-spectral data are transmitted to the solar activity forecast center at NOAA in Boulder, Colorado where the information is used for operational forecasting. Beginning in late 1975, the observing site served as a facility for testing a prototype of an elaborate radio-spectrograph/polarimeter that will be launched in 1977 and which is scheduled to fly by Jupiter in 1979 and Saturn in 1981.

SPACE ENVIRONMENT LABORATORY. NOAA, Boulder, Colo. 80302. Both optical and radio measurements of solar activity are made on a daily basis at SEL. The optical instruments, located at 325 South Broadway, consist of a 5-inch refractor (utilizing a Halle H-alpha filter) for viewing the solar chromosphere; a 4-inch refractor for viewing the sun in white light; a 9-inch heliostat for spectroscopic studies; and an auxiliary 12-inch heliostat. The optical observatory conducts a routine full-time solar flare patrol and provides the Forecast Center with a daily H-alpha photograph, a daily sunspot drawing and sunspot polarities. The solar radio monitors are located at the Table Mountain NOAA facility some 9 miles north of town. They presently consist of a 245 MHz system using two 12-element yagis, a 606 MHz system using four 15-element yagis, radiometers at 1465, 2995, and 4995 MHz using a single 2-meter parabolic dish, and an 8800 MHz radiometer using another 2-meter parabolic dish antenna. These radio systems are operated during daylight hours to provide the Forecast Center with solar radio burst and noise storm information.

Connecticut

TALCOTT MOUNTAIN SCIENCE CENTER, INC. Montevideo Rd., Avon, Conn. 06001. 1.5 mi. from Rt. 44 at Avon-West Hartford border; private road, entrance by appointment. Visits and telescopic observation by arrangement. Write for current information; the Center runs 8-session programs with a tuition fee of $30 for members, $35 for others. All sorts of group visits and field trips are arranged with and for school groups. Telescopes for visual observing are the 12-inch Tinsley Cassegrain, an 8-inch Newtonian, an 8-inch Schmidt-Cassegrain, seven 6-inch Newtonians, and four refractors, three of which are for solar use. A 5-inch refractor here was used to discover two comets. The Center provides equipment necessary for astrophotography. Also available is a 1.5-meter Wadsworth spectrograph and a 120 mm coelostat. Speakers on astronomical subjects can be obtained at the same rate as at the center to address groups elsewhere.

WESTERN CONNECTICUT STATE COLLEGE OBSERVATORY. 181 White St., Danbury, Conn. 06810. At Higgins Sciences Bldg. No scheduled hours; open to school groups as requested. Permanently mounted equipment consists of a large Schmidt-Cassegrain and a 6-inch refractor. There are also two portable instruments: an 8-inch Newtonian and a Questar. These serve for teaching plus some photometry of variable stars. BA, BS, and MS offered here in earth and space sciences.

JOSEPH HALL OBSERVATORY. Hartford Public High School, 55 Forest St., Hartford, Conn. 06105. Public viewing Tue and Thur nights, weather permitting, beginning at 7 p.m. in winter and 9 in summer, with a planetarium show preceding observations. The instrument is a 9½-inch Alvan Clark achromat of 135-inch focal length on a Warner and Swasey mount; tracking is by a weight-driven clock with centrifugal governor, and the mount is equipped with very finely divided setting circles. According to D. J. Warner, a Hartmann test made in 1930 showed the lens of this instrument to be of exceptional quality. Formerly overlooking the State Capitol, the observatory was moved to its present location in 1963. It is used for teaching and public observation. Joseph Hall obtained this Observatory for Hartford Public High School in 1884; he was then principal of the school and distantly related by marriage to the Clarks. This telescope provides residents of a densely populated area the chance to see the moon and planets with unusual clarity.

VAN VLECK OBSERVATORY. Wesleyan University, Middletown, Conn. 06457. On a hill on the western part of the campus. Open house at irregular intervals. Wesleyan's 20-inch f/16.5 refractor is the largest in New England; it was made by the Alvan Clark firm in the early 1920s—though by that time all the Clarks were long dead, and even Carl Lundin lived only until 1915. The optical quality is, however, excellent; a double-slide plate carrier made in a machine shop at Wesleyan has made possible the sharp photography necessary for astrometric work (see discussions of Allegheny Observatory and the Flagstaff Naval Observatory in Part I). A recent development at Van Vleck in this field has been the use of computers to allow for optically induced error and to search for statistical correlations. The 20-inch is still used for determining parallaxes and proper motions of nearby stars, most recently in a program involving late-type dwarf stars; twenty such stars in the Hyades cluster are being studied. Photometry programs carried out at Kitt Peak are correlated with these measurements. Van Vleck has recently added a 24-inch f/13.5 Cassegrain of its own for photometric work (and also for photography). By far the most interesting piece of equipment here, from a historical point of view, is still doing service for student observing; this is a 6-inch refractor called the "Fisk" telescope—not for the maker but for the purchaser, President Wilbur Fisk of Wesleyan who bought it in 1836 from the Paris firm of Lerebours. Until Cincinnati and Harvard acquired larger lenses about ten years later (q.v.), this was the largest telescope in the country. Housed in various places for more than thirty years, the 7-foot-long brass telescope with its teakwood stand was finally supplanted by a 12-inch Alvan Clark made in 1869. In 1959 it was brought out of retirement, and now is housed in a fiberglass dome on the west end of the Observatory. The 12-inch, by the way, was given to Leslie Peltier because of his activities as a comet-searcher and a variable star observer; it is now in Delphos, Ohio, doing good duty. Since Van Vleck—or, at any rate, Wesleyan—has been in business in astronomy since the beginnings of that science in this country, there is a quantity of auxiliary equipment and small instruments of historical interest around the Observatory: transit instruments, a 5-inch finder for the 20-inch, a filar micrometer, spectroscopes, other small portable telescopes, and various kinds of cameras. The active astronomy program offers both BA and MA in astronomy, and encourages student research.

CENTRAL CONNECTICUT STATE COLLEGE PLANETARIUM AND OBSERVATORY. New Britain, Conn. 06050. In Copernicus Hall. The 117-seat Spitz 512 planetarium is open to the public during the academic year on specified dates and by arrangement; there is also a cooperative program with the New Britain School System. The Observatory is available to qualified outside groups for public nights. Observatory equipment includes a 16-inch Group 128 Cassegrain, used for photography; two 8-inch Celestrons for visual and photographic work; and a 4-inch Clark refractor that is used for visual work. BA, BS, and MS degrees in physics and earth science are offered, with considerable course options in astronomy. This is a well-balanced instructional facility.

YALE UNIVERSITY OBSERVATORY. Dept. of Astronomy, Watson Astronomy Center, P.O. Box 2003 Yale Station, New Haven, Conn. An 8-inch refractor on campus is used for teaching and for irregularly scheduled lectures and observing sessions for the New Haven Astronomical Society. Located in Bethany, Conn., is a 20-inch reflector and a 9-inch Clark refractor, along with several smaller instruments. The 20-inch is used for photometry and spectroscopy. Yale has a 40-inch reflector in Chile, at Cerro Tololo, which is used for astrometry, photometry, and spectroscopy; at El Leoncito in Argentina it operates a 20-inch double astrograph. Yale was one of the earliest institutions to own a good refracting telescope in this country; a 5-inch Dollond arrived in 1830, but was not well mounted and was used through the windows of an octagonal tower on a campus building. The independent recovery of Halley's Comet in 1835, however, by Elias Loomis of the College, made quite an impression (see Allegheny and Harvard in Part I for accounts of other benefits bestowed by comets on observatories). By 1882 there was an Observatory building, and a heliometer, for measuring the sun's diameter, was purchased from the Repsolds and set up in the west dome. It turned out that this instrument was ideal for visual parallax work, and with measurements of the positions of the Pleiades, William Elkin showed that most of the stars were associated in a cluster. At one time or another and in various locations, Yale has owned and used an 8-inch Grubb refractor, a 26-inch photographic refractor by Brashear, the 15-inch Loomis Polar refractor, Ross cameras, and Clark refractors. It would be interesting to know where they are, or were, and which are still in use or on display. The current teaching and research staff in astronomy at Yale includes more than a dozen astronomers, who work in most of the special fields. The main strength of the department lies in stellar and galactic astrophysics, astrometry, cosmology, and relativistic astrophysics. Recent theoretical research topics have included the study of star formation, the internal structure and atmospheric properties of horizontal branch stars, thermal instabilities in stellar interiors and associated problems of nucleosynthesis, the effects of rotation on stellar evolution, and black holes. Observational projects have included studies of star clusters with regard to stellar evolution, determination of absolute stellar proper motions, period changes in variables, and light distribution in elliptical galaxies. Departmental headquarters are located in the Watson Astronomy Center, which also houses a library and measurement laboratories; equipment in the latter includes an iris photometer, a Grant spectrum comparator, a blink microscope, a Mann measuring engine and a microdensitometer.

STAMFORD OBSERVATORY and EDGERTON MEMORIAL PLANETARIUM.
Stamford Museum and Nature Center, 39 Scofieldtown Rd., Stamford, Conn. 06903. Located 3/4 mile north of Merrit Parkway on High Ridge Rd., at intersection with Scofieldtown. Free visits to the Observatory 8–10 p.m. Fri and by arrangement for groups. There is a 22-inch Maksutov that is used for variable star work and positional astronomy in observing comets, in cooperation with various national organizations. This center is extremely active in promoting amateur and popular astronomy; on the first Fri each month the Fairfield County Astronomical Society meets to hear a speaker or watch a film; new members and casual visitors are welcome. A Junior division meets the second Fri at 8:30 p.m. for a similar program for younger visitors. A live radio program, "Window to the Sky," is broadcast over a local AM station Thur 7:30–8:30 p.m. Several evenings each week there are amateur telescope-making classes, for which one may register. Amateurs may use their own instruments on an observing deck equipped with electrical outlets. It is a first-rate amateur center, and one of the largest telescopes for public viewing in the country. The Planetarium has shows at 3 p.m. Sun; admission: adult $.75, child $.50. Group visits by appointment. It has a 23-foot dome seating 55, and uses a

Spitz A-1 projector. At the center there is also a bookstore and gift counter, and there are numerous scientific displays and exhibits.

Other observational telescopes:
Eastern Connecticut State College
Loomis Institute

Mystic Seaport P., Mystic
Southern Connecticut State College,
   New Haven
Hillcrest Junior High School, Trumbull

Larger planetariums:
Museum of Art, Science, and Industry,
   Bridgeport

Delaware

MOUNT CUBA ASTRONOMICAL OBSERVATORY, INC. P.O. Box 3915, Greenville, Hillside Mill Road, Wilmington, Del. 19807. Open evenings of first and third Mon each month. Admission by ticket only: free, but write in advance. Special group visits can be arranged. Leaflet with map available. This Observatory is unique in that it is a self-sustaining nonprofit research center and educational facility, with ties to the University of Delaware. It has its own Board of Trustees to whom persons wishing to carry out research projects may submit their proposals. The main instrument resembles that of New Mexico State's: a 24-inch Cassegrain with long and extra-long focal length secondaries (32 and 64 feet). It is used for photoelectric photometry and determination of asteroid positions. A Baker correction system makes possible 5-degree photographic plates. All optics are quartz. In one wing of the building is a lecture room, a planetarium, and a 4½-inch refractor (made originally by Brashear in 1887 with an Alvan Clark objective). These latter facilities are used for the public nights, which consist of a talk, a planetarium show, and an observing session. In bad weather, an additional session of varying character is held. Supporting facilities include a shop, a photographic dark room and an astronomical library. On the same grounds is the **OBSERVATORY OF THE DELAWARE ASTRONOMICAL SOCIETY** (P.O. Box 652, Wilmington, Del. 19899) which has a 12½-inch reflector in a sliding roof observatory.

District of Columbia

(see also entries under Maryland)

MUSEUM OF HISTORY AND TECHNOLOGY. The Smithsonian Institution. On the Mall in Washington, D.C. In addition to many things found in other museums, such as a Foucault pendulum and demonstrations in technology of all sorts, this contains what may be the best collection of astronomical optics of historical interest in the world. This includes a reproduction of Henry Fitz's optical shop, and lenses and mirrors with which many important discoveries were made. Rivaled only by the Adler Planetarium's museum in this country, and the London Museum abroad.

NATIONAL AIR AND SPACE MUSEUM. On the Mall in Washington at 7th St. and Independence Ave. In addition to the exhibits of aircraft, spacecraft, rockets, and satellites, this museum's new building will contain a Zeiss Model VI planetarium projector under a dome seating 250.

U.S. NAVAL RESEARCH LABORATORY. Washington, D.C. 20390. In Washington the Navy has a 15.2-meter parabolic antenna. For other installations in and around

Washington, see the long article on the Naval Observatory in Part I, and the entries in the Maryland section of this catalog.

Other observational telescopes:
George Washington University
Howard University

Larger planetariums:
Rock Creek Nature Center

Florida

UNIVERSITY OF MIAMI. Physics Dept., P.O. Box 8284, Coral Gables, Fla. 33124. Portable instruments are used for instruction, and are set up on the roof of the administration building. One member of the Physics Dept. has studied coronal emission lines with ultrahigh dispersion during eclipses. Several courses in astronomy, and a joint physics-astronomy major are given.

BROWARD COMMUNITY COLLEGE OBSERVATORY. 3501 Davie Rd., Fort Lauderdale, Fla. 33314. Open Thurs nights for the general public. The 12-inch Cassegrain and 6-inch refractor are mainly used for teaching and visual observation, though the Cassegrain is used for some photometric work. The College also operates the Buehler Planetarium.

OPTICAL OBSERVATORIES OF THE UNIVERSITY OF FLORIDA. Dept. of Physics and Astronomy, Gainesville, Fla. 32611. Tours only upon special arrangement. The teaching Observatory has an 8-inch refractor and a 5-inch reflector, for instruction and student projects. **ROSEMARY HILL OBSERVATORY** has a 30-inch reflector for photographic and photoelectric photometry. Quasars, eclipsing binaries, and flare stars are the objects of these measurements. Other instruments include 18- and 12½-inch reflectors used for the same types of programs.

JOHN YOUNG MUSEUM AND PLANETARIUM. 810 East Rollins St., Orlando, Fla., 32803. "Regularly scheduled 'skywatch' sessions are conducted for the general public using our 12½-inch Cave reflector, a 9-inch f/8, a 6-inch f/8 and a 6-inch RFT (f/4). On the roof of the new museum wing, just completed, we hope to mount our instruments under a dome. After Jan. 1, 1976, there will be a 50¢ admission charge to the Museum, including the observatory. The planetarium curator and observatory director, Bruce Salmon, carries out visual atmospheric and albedo studies of the planet Mars."

U.S. NAVAL OBSERVATORY TIME SERVICE SUBSTATION. Richmond, Fla. The U.S. Postal Service returns mail addressed to Richmond, claiming no such location; it is, however, a few miles west of Miami in southern Florida. Experience at other branches of the Naval Observatory would indicate that visits are possible. A photographic zenith tube (see Naval Observatory—Washington, in Part I for details on its operation) was installed here in 1949. The more southerly latitude permits use of different stars for zenith timings.

SATELLITE BEACH OBSERVATORY. Satellite High School, Brevard County Schools, Scorpion Lane, Satellite Beach, Fla. 32937. "Open upon reasonable request—classes have access." The largest piece of equipment is a 16-inch f/6 Cave reflector, but there is also a 5-inch f/2.5 astrograph, an 8-inch f/4 reflector for comet seeking, and a 23-camera meteor patrol. Classes have assembled 36 portable f/10 Newtonians that are used for observing grazing occultations. The Observatory is "very heavy on grazing occulta-tions" in its lunar occultation work, and participates in the Florida Fireball Network for tracking large meteors. The director points out that at 28° 12′ N. Latitude, they have a good opportunity for making first U.S. confirmation of a southern nova or comet. The research has borne fruit with the discovery of 9 new binaries during the observation of 131 grazing occultations, and 27 solutions for fireball orbits. This is an example of a high school observatory which, by the disciplined pursuit of feasible goals, has accomplished some really valuable research—far more than many a university observatory.

UNIVERSITY OF SOUTH FLORIDA OBSERVATORY. Dept. of Astronomy, Tampa, Fla. 33620. On 46th St., ½ mile north of north edge of the campus (Fletcher Ave.). Public nights monthly and on special occasions, except in summer. No fixed schedule. The main telescope is a Tinsley 26-inch Schmidt-Cassegrain (1968) with a 400-inch focal length. Equipment for this consists of a Boller and Chivens grating spectrograph, a 2-channel photometer, and an astrometric camera; and a hydrogen-alpha monochromator with an image-intensifier tube. There are also two 12-inch Cassegrains of 170-inch focal length, and a photometer; and the observatory owns a Mann measuring engine. The main telescope was the first one of such size to use the Cervit material now selected for the Cerro Tololo mirror. It was used for very accurate observation of the occultation of β Scorpii by Io in May, 1971. Research specialties include astrometry, image tube spectroscopy, and eclipsing binary photometry. A major use is also for instruction. The Dept. of Astronomy offers BA and MA degrees and cooperates with the University of Florida in a Ph.D. program. There is a good planetarium that is also part of the Astronomy Dept.

Other observational telescopes:
Edison Community College
Florida Junior College
Florida Southern College
Florida Technological University
Pensacola Junior College
Santa Fe Community College
University of Western Florida

Larger planetariums:
Bishop P., South Florida Museum,
   Bradenton
Jacksonville Children's Museum
Miami Museum of Science
E. G. Owens P., Pensacola Junior
   College, Pensacola
Pensacola Naval Air Station
Science Museum and Planetarium of
   Palm Beach County

Georgia

FERNBANK SCIENCE CENTER OBSERVATORY. Dekalb County Board of Education, 156 Heaton Park Dr., Atlanta, Ga. 30307. On the east side of Atlanta, near Decatur. Public hours every Fri evening; public courses are also offered. Fernbank boasts a research-grade telescope: a 36-inch Tinsley Cassegrain with auxiliary equipment that includes a medium-dispersion spectrograph, a 4 x 5-inch plateholder, a photo-electric photometer, a cool emulsion camera, and a TV image system. Research is carried out in photoelectric photometry of eclipsing binaries plus a photographic patrol of BL Lacertid objects. The TV image system has tracked several Apollo Space Craft on their way to the moon. For visual observing there are two 5-inch Celestrons, an 8-inch, and a 10-inch, plus two 6-inch Cave Astrola reflectors and a 4-inch Cave refractor. For photography there is an 8-inch Celestron Schmidt camera. This center, located in a large virgin forest between Atlanta and Decatur, has been in operation since 1967. It rivals the

Morehead Planetarium in North Carolina as the best multipurpose astronomical center in the South. It is one of the half-dozen best in the country.

BRADLEY OBSERVATORY. Agnes Scott College, Decatur, Ga. 30030. Inquire about current public hours. The Observatory owns a 30-inch Cassegrain, one of the largest telescopes owned by any college of this size. It is used for teaching and is available to advanced students and amateurs for research projects. The telescope is housed in an attractive building on the campus.

MUSEUM OBSERVATORY. Museum of Arts and Sciences, 4182 Forsyth Rd., Macon, Ga. 31204. On U.S. 41 N in north Macon. Public viewing without charge after planetarium shows on Fri, at 9:15 p.m. A Celestron 14 is on order, and the Observatory already has a 10-inch and two 8-inch Celestrons. All of these are used for popular astronomy and in connection with high school classes. The **MARK SMITH PLANETARIUM** in the same building has a 40-foot dome and seats 200; write for a current schedule or to arrange group shows. For purposes of public astronomical education, this is one of the best centers in the South.

VALDOSTA STATE COLLEGE OBSERVATORY. Valdosta, Ga. 31601. Viewing by groups upon request, often in connection with a show in the planetarium, which is equipped with a Spitz A3-P projector. Get in touch with the Dept. of Physics and Astronomy. Special shows for eclipses, comet apparitions, and so on. A $12\frac{1}{2}$-inch Cassegrain is equipped with a photoelectric photometer, a camera, a spectrograph, and a filar micrometer. There is also a 10-inch Newtonian, a 4-inch astrographic camera, and a large collection of small reflecting and refracting telescopes. The college offers a BS in astronomy as well as in physics.

Other observational telescopes:
Emory University
Georgia Institute of Technology
Georgia Museum of Arts and Sciences
Georgia State University

Larger planetariums:
Fulton High School, Atlanta
Northside High School, Atlanta
Walker County Space Science Center,
  Rock Spring
Savannah Science Museum

Hawaii

MAUNA KEA OBSERVATORY. Institute for Astronomy, 2680 Woodlawn Dr., Honolulu, Hawaii 96822. Observatory office: 180 Kinoole St., Hilo, Hawaii 96720 Telephone (808) 935-3371. SR 20 (Saddle Rd.) leads to turnoff at Humuula Sheep Station and road to the Observatory. Open daily to visitors 9:30 a.m. to 3:30 p.m. except weekends and holidays. During spring and summer months, open houses at night are scheduled monthly around the full moon. The main instrument is the University of Hawaii's 88-inch reflector which can be used in either Cassegrain or coudé configuration. Thus far, it has been used for spectroscopy, photometry, infrared radiometry, photography, polarimetry, and interferometry, where the mirrors are arranged so as to send the light through the coudé aperture; the beam ends up in a room below the level of the telescope after being reflected five times over a path of about 250 feet. Two Boller and Chivens 24-inch Cassegrain telescopes serve very diverse purposes—one being employed for many of the same programs as the 88-inch telescope, and the other being entirely devoted to the international Planetary Patrol (see Lowell Observatory in Part I). A

Canada-France-Hawaii 144-inch reflector telescope is scheduled to be put in operation at Mauna Kea in 1978. The building for the $22 million facility is under construction at the present time. A 3-meter, $6 million, Infrared Telescope Facility funded by the National Aeronautics and Space Administration to be built on the summit of Mauna Kea, should be operational in 1977; construction will begin in the spring of 1976 on a $3 million, 3.8-meter United Kingdom Infrared Telescope which is expected to be operational by 1978. For those who can get there, a visit to the Mauna Kea Observatory must rank among the most spectacular. Although there is a paved road to the 9,200-foot level, the slope of the last 6 miles is such that it can only be traversed by four-wheel-drive vehicles. In case of heavy snow, which usually occurs two or three times each winter, a tracked vehicle is used for access. Inquire at the Hilo Institute for Astronomy office about conditions before setting out. The 13,800-foot summit of this dormant volcano is more than twice the elevation of any major nighttime observatory in the United States; it is an example of a "shield" volcano—one whose summit is built up by innumerable rounded layers of lava flows. The latitude of the Observatory is also of interest: at 19°40′ N. latitude, it is more than 200 miles closer to the equator than Key West, and hence its telescopes have better access towards the center of our galaxy than any in the continental United States. Vegetation ceases at about 10,000 feet, above the barren surface of which rise several small cinder cones. The atmosphere is extremely dry, the proportion of clear nights among the highest in the world, and pollution from smoke, dust, and streetlights at a minimum. It will be chilly at any time, but proximity to the equator has kept the lowest recorded temperature to about 10 degrees F; mean minimum is close to 30 degrees F, summer and winter. Summer daytime temperatures may reach 50 degrees F. The altitude, comparable to that of Mount Evans in Colorado, may make those who drive directly up from sea level light-headed, and may cause headaches or make some people nauseated. Visiting observers are required to spend a day getting acclimatized, and to sleep at a mid-elevation facility about 9,000 feet. There are several state parks in Hawaii, including Pohakuloa State Park, and some tent sites in the Volcanoes National Park. In Hilo, the Hilo Hukilau Hotel and the Hilo Hotel have had single rooms available for $12 and doubles for $15, but these rates may fluctuate.

An improved road to the summit of Mauna Kea was under construction in 1976.

HALEAKALA OBSERVATORIES. University of Hawaii, P.O. Box 135, Kula, Maui, Hawaii 96790. Located in Haleakala National Park near the summit area in a place called Kolekole. A paved road brings the visitor to the summit from Kahului Airport in about an hour and a half. Inquire about arranging a visit. Located on the 10,000-foot summit are the C.E.K. Mees Solar Observatory, a Zodiacal Light Observatory, and the Lunar Ranging Observatory. NASA, the National Science Foundation, and the University of Hawaii all support programs under way here. At the Mees Observatory, a 10-inch coronograph and a coudé spectrograph are used for studies of the inner corona, coronal condensations, and prominences. To quote verbatim the pamphlet on Haleakala: "The f/30 coronograph gives an image scale of 33 arcsec/mm at the coudé focus. Two interchangeable diffraction gratings provide dispersions of 5.5 and 1.0 A/mm; the spectrograph can be used in either single- or double-pass configuration." There is also a polarimeter-photometer with a 6-inch objective, which automatically scans portions of the sun and stores information about its light-polarization in a computer. A Czerny-Turner echelle spectrometer has been recently added to this. There is a slow-scanning coronal spectrophotometer that uses a thermoelectrically-cooled silicon-vidicon camera tube. Other telescopes monitor the sun using narrow bandpass filters; these include a dual coronograph and hydrogen-alpha and K-line patrol telescopes. There is also a flare patrol telescope. (For a description of other solar equipment, see Sacramento Peak,

N.M., and Big Bear, Cal.) Haleakala is certainly appropriately located; the name means "House of the Sun." Also, though, the Zodiacal Light Observatory studies the light scattered by interplanetary dust; mainly this is done by scanning the skies for polarized light that can be attributed to the dust. The 5.75-objective housed in a sort of silo does this, at a rate of a degree or two a second. The Lunar Ranging Observatory uses a laser beam that is expanded and recollimated with a 16-inch telescope, then aimed by a flat at one of the retro-reflectors on the moon. (See McDonald for similar work.) Eighty separate 8-inch lenses receive the returning light and detection is with a photomultiplier tube. Perhaps this is the "90-inch equivalent" telescope developed by the Joint Institute for Laboratory Astrophysics (Colorado) for such work. At any rate, the earth-moon distance will be measured to within inches.

PALEHUA OBSERVATORY. Air Force Cambridge Research Lab, Hansom Field, Bedford, Mass. 01730. Located at Oahu, Hawaii. A 2.44-meter paraboloid monitors the sun's radio emissions.

KAUAI OBSERVATORY. Radio Astronomy Branch, Goddard Space Flight Center, Greenbelt, Md. 20771. Located at Kauai, Hawaii. A 5-element Yagi interferometer monitors Jupiter.

Other observational telescopes:	**Larger planetariums:**
Hilo College	Berenice P. Bishop Museum, Honolulu
Leeward Community College	

Idaho

HERRETT ARTS AND SCIENCE CENTER. East Five Points, Twin Falls, Idaho 83301. Write or call (208) 733-0868 for group tours. This is an extensive museum-classroom-laboratory complex set up by Norman Herrett for the education and entertainment of children; it has recently become affiliated with the College of Southern Idaho. High school and college students by the score participate in museum lectures, planetarium demonstrations, and so on. Astronomical equipment is a 10-inch catadioptric telescope with an 18-foot focal length, and a 12½-inch Newtonian, in addition to the planetarium.

Observational telescopes:	**Larger planetariums:**
Boise State College	Ricks College, Rexburg
College of Idaho	
Idaho State University	
University of Idaho	

Illinois

MARK EVANS OBSERVATORY. Illinois Wesleyan University, Bloomington, Ill. 61701. At Park and Beecher Sts. Public nights about once a month; write for schedule. The 16-inch Ealing Cassegrain has a 5-inch Celestron guide scope, a wide field f/2.5 camera, and other accessories. There is a 21-foot Jarrel-Ash spectrograph and a photoelectric photometer. Work is in progress on radio equipment. Specialized investigations are carried out in instrumentation and solar spectroscopy by the staff and advanced physics students; a BA in physics is offered. R. Wilson writes: "The Mark Evans Observatory was built on the site of the Behr Observatory. The Behr Observatory was built in 1895 and housed an 18½-inch Newtonian telescope built by John Calver of England. We understand that in 1895 it was the eighth largest telescope in the U.S. We still have some eyepieces and the main mirror, scribed on the back by Calver. Also on campus, 4-inch and 6-inch lenses of European make, quite old, and a sextant made by Rennison and Sons, England. It is our belief that the 18½-inch Newtonian had seen service at Yerkes or in Chicago. It has been used by Professors Burnham, Hale, and Hough."

ADLER PLANETARIUM. 900 E. Achsah Bond Dr., Chicago, Ill. 60605. Located on the peninsula in Lake Michigan near the Field Museum, reached from the north-bound lane of Lake Shore Drive. Open Jun 16–Aug 31 daily 9:30 a.m. to 9 p.m.; rest of year Mon–Fri 9:30 a.m.–4:30 p.m. and Sat, Sun, and holidays to 5 p.m., Tue and Fri to 9:30 p.m. Closed on Christmas, New Year's Day, Thanksgiving. Daily Sky Shows in winter at 2 p.m. and at 11 a.m., 1, 3, and 4 p.m. weekends and holidays; six shows per day in summer; 7:30 p.m. shows Tue and Fri all year. Admission $1 adult; $.50 under 18; under age 6 not admitted. Free admission to the museum, which is actually the best part. The Planetarium is a fine one, with a Zeiss VI projector, many special effects, and a seating capacity of 392 under a 68-foot dome. But the museum is unique; except for the London Science Museum and the Museum of History and Technology in Washington, there is nothing to compare it with. Some memorable features include a glass-fronted telescope-making shop, where amateurs can grind their own mirrors (and be watched by the public during open hours); a working coelostat and solar telescope to show sun-spots—working, that is, except when the sun is too low in winter; extensive exhibits of ancient astrolabes and time-keeping apparatus; the original wooden tube of the Clark 18½-inch refractor owned by Dearborn Observatory (q.v.); originals and models of early telescopes of all types; excellent displays and demonstrations about stellar evolution, cosmology, and the solar system; and a fine shop and bookstore. The Field Museum down the street has meteorite and geology exhibits. (See "Nearby" under Yerkes in Part I for other suggestions.)

DEARBORN OBSERVATORY. Northwestern University, Evanston, Ill. 60201. At 2131 Sheridan Rd., which runs north along Lake Michigan, as a sort of continuation of Chicago's Lake Shore Drive, terminating within the campus. Visits: Apr–Oct Fri evenings, two visits: 9–10 and 10–11 p.m. Free tickets must be obtained in advance by calling (312) 492-7651; calls for tickets not accepted before second week of March. The 18½-inch Clark refractor is one of the most impressive instruments available for public viewing. This lens had been ordered by the University of Mississippi, but the outbreak of the Civil War resulted in its being sold in 1863 to a group from Chicago who outbid the Harvard Observatory. Alvan Clark was testing it when, from behind a wall in Cam-bridge where he had it set up, popped the hitherto unseen companion to Sirius, a few seconds before the bright star itself came into view; this then accounted for the oscilla-tions in Sirius's proper motion. The 18½-inch is equipped with a plate camera, a spectro-graph, and a Filar Micrometer as well as the original Clark eyepieces. It has been used to pioneer image Orthicon techniques. Its original wooden tube is downtown in the Adler Planetarium, whose several floors of displays are perhaps the best in the western hemi-sphere. (See also the description of the Lindheimer Center, on the Northwestern Campus.)

LINDHEIMER ASTRONOMICAL RESEARCH CENTER. Northwestern Uni-versity, Evanston, Ill. 60201. On Lake Michigan at extreme eastern edge of Northwestern University Campus. Visits every Sat 2–4 p.m. This remarkable structure is worth visiting for its architecture alone—a great exterior supporting frame of semigeodesic design and the two domes perched atop towers joined by a sort of flying bridge between. It rivals the Church of the Segrada Familia in Barcelona for pure fantasy. One may see the 40-inch and 16-inch reflectors. At its Cassegrain focus, the 40-inch has a spectrograph, a rapid-scan spectrophotometer, and a plate camera that may be attached. Faculty and graduate students use this instrument, while undergraduates use the 16-inch. Studies are carried out on the spectroscopy of close binaries, spectrophotometry of emission stars, and quasar variability. (For telescopic observation by the public, and more complete travel

Lindheimer Astronomical Research Center, near Lake Michigan, an example of imaginative astronomical architecture. One telescope is visible through the slit. (Photo: courtesy Northwestern University)

directions, see Dearborn Observatory.) The campus lake fill south of the Observatory is favorable for amateur observation.

WALTER H. BALCKE OBSERVATORY. Illinois College, Jacksonville, Ill. 62650. At Crispin Science Hall. Public hours available for any group on request. The 10- and 14-inch Celestrons serve for teaching and public observation here at the oldest institution of higher learning in Illinois.

ILLINOIS BENEDICTINE COLLEGE OBSERVATORY. I. B. C. Astronomical Society, 5700 College Rd., Lisle, Ill. 60532. At the corner of College and Maple Rds. Open to members, which consist of interested and responsible persons. The Society believes its 16½-inch reflector on a Springfield mount to be the largest of this type; the Springfield arrangement is in some ways comparable to the coudé and allows an observer to sit motionless at the eyepiece while the telescope moves. There is also a 5-inch alt-azimuth refractor. In planning stages are a pair of small paraboloid radio dishes to be used as an interferometer. An account of the Observatory in *Sky and Telescope* for December 1962 reported with astonishing matter-of-factness that: "After considering many telescope designs, Edmund Jurica, O.S.B., drew up plans for a 16½-inch, f/7.5 Springfield reflector and a domed observatory. Construction of both was undertaken by Brother Andrew Havlik, director of our physical plant." Such a project might be a challenge for a long-established optical firm.

NORTHMOOR OBSERVATORY. Northmoor Golfcourse, Peoria, Ill. 61614. Open Sat evenings from sunset to 10 p.m. May–Oct. Adults 12 and over, $.25. There is a 9-inch refractor and a 12-inch reflector.

CARL GAMBLE OBSERVATORY and JOHN DEERE PLANETARIUM. Augustana College, Rock Island, Ill. 61201. On campus at 38th St. and 7th Ave. Visits by school groups, church groups, clubs, and the general public may be arranged; inquire locally. A 6-inch Zeiss refractor is housed on top of an imaginatively designed tower that mushrooms out to an observing deck some 30 feet above the ground. A Spitz A3P serves the planetarium. Displays include an extensive meteorite collection, transparencies of celestial objects, meteoroid detectors, and materials related to the U.S. Space Program. An unusual feature is a 556-pound iron meteorite from the Barringer crater in Arizona.

PRAIRIE OBSERVATORY. University of Illinois, Urbana, Ill. Located at Oakland, 30 miles south of Urbana. No public visits. The 40-inch Cassegrain/coudé is used with photometric and spectrographic equipment. At one time the University of Illinois ordered a 12-inch refractor from John Brashear. Very little information was available from this astronomy department.

VERMILION RIVER OBSERVATORY. University of Illinois, Urbana, Ill. 61081. Located near Danville, Ill. The big 36.6-meter dish is mounted equatorially, and is used to observe both the lines and the continuum of galactic and extragalactic radio sources.

Other Observational telescopes:	Larger Planetariums:
Elgin Community College	Elgin Observatory and Planetarium,
Illinois College	Elgin
Illinois State University	Reed P., Olivet Nazarene College,
Lincoln Land Community College	Kankakee
Loyola University	Illinois State University, Normal
Wheaton College	

Indiana

HAROLD E. ROZELLE OBSERVATORY. Anderson College, Anderson, Ind. 46011. At 1100 E. 5th St. Open Tue and Thur from dark until 11 p.m. A 14-inch Celestron with a 5-inch guidescope is being installed in a dome; there are also reflectors of 6 and 4 inches. Research programs are not yet planned.

KIRKWOOD OBSERVATORY. Indiana University, Bloomington, Ind. 47401. Open every clear Wed evening except during vacation periods. A 12-inch aperture telescope is used for teaching and observation. At one time this observatory owned a 12-inch Brashear refractor. Morgan Monroe State Forest offers a possible location for amateur camping and observation.

GOETHE LINK OBSERVATORY. Indiana University, Brooklyn, Ind. 46111. Admission to public nights by free ticket only; send a stamped, self-addressed envelope. Three open nights per year, announced as scheduled. This observatory specializes in photoelectric photometry and spectrophotometry, using three principal instruments: a 36-inch reflector, a 16-inch reflector, and a 10-inch photographic reflector.

KOCH PLANETARIUM. Evansville Museum of Arts and Sciences, Evansville, Ind. 47713. Free public shows the first Sat each month.

FOX ISLAND OBSERVATORY. Fort Wayne Astronomical Society, P.O. Box 6004, Fort Wayne, Ind. 46806. In the Fox Island Nature Preserve, Yonne Rd. Public hours to be announced when the new dome is in place in 1976; presently the Observatory is located on the north side of Fort Wayne and has been open Sat nights. The main instrument is a 12½-inch reflector; each member has attachments, such as cameras, solar filters, a photometer, and spectral prisms. A very active astronomical society, some of whose members have taken advanced degrees and become professional astronomers. Their observatory is presently called Mount Willig Observatory.

McKIM OBSERVATORY. DePauw University, Greencastle, Ind. 46135. At Highridge and DePauw Aves. No public hours. There is an Alvan Clark refractor with a 9.53-inch aperture on a Warner and Swasey mounting. The Observatory also owns several items now of great historical interest, but which were originally just part of its general equipment. Most important of these is a transit telescope, but there are also cameras and chronographs. Interest seems to be reviving in the Observatory at DePauw, and one cannot refrain from remarking that the neglect and disuse of the Observatory can be turned to very good account. Since it has not been continually modernized, it seems to furnish the outstanding example in the country of an excellent nineteenth-century observatory. One wishes that the Smithsonian could arrange to have it preserved and regularly open as a museum piece. Since such optics and machinery undergo little deterioration, the restorations undertaken so far have been most successful. At the time that it was actually built, with money given by the wealthy contractor, Robert McKim, such a fine facility was really much beyond the real needs of the University. With five rooms, including a library, a transit room, a chronograph room, a clock room, and an equatorial room, it was a kind of Greenwich-in-the-Mid-West. Used for visual observing, such instruments as the Clark refractor are superior to any other type of telescope for views of the planets and the moon under adverse conditions of light and air pollution. This is one of the most interesting small observatories in the world from an historical point of view.

J. I. HOLCOMB OBSERVATORY AND PLANETARIUM. Butler University, Indianapolis, Ind. 46208. Open 3–5 and 7–9 p.m. Sat and Sun without charge; by reservation on weekdays. The 6-inch refractor and the 38-inch Cassegrain are used for teaching and for various kinds of research.

INDIANA STATE UNIVERSITY. Terre Haute, Ind. 47809. Atop the Science Bldg. at 6th and Chestnut Sts. Public nights every Tue and for certain astronomical events. Classes and groups by arrangement. Telescopes up to 8-inch aperture available for observing.

VALPARAISO UNIVERSITY PLANETARIUM. Valparaiso, Ind. 46383. Located in the Neils Science Center on Campus. The Viewlex Apollo projector is used for shows at 3 p.m. the 2nd and 4th Sun of each month, and by appointment.

Other observational telescopes:	Larger planetariums:
Anderson College	Virgil Grissom P., Arlington High
Ancilla College	School, Indianapolis
Indiana Central College	Ball State University, Muncie
Purdue University	Hamilton Southeast High School,
St. Francis College	Noblesville
University of Evansville	Dunseth P., Vincennes University,
	Vincennes

Iowa

ERWIN F. FICK OBSERVATORY. Iowa State University, Ames, Iowa 50010. Location and public hours not supplied. The principal instrument here is the Mather Telescope, a 24-inch Cassegrain which has recently been converted for work at coudé focus. For this purpose a spectrometer has been acquired. Various cameras, including one with an image tube, can be used. There is a rapid-spectrum scanner that was used in studies of comet Kohoutek. Astronomers and graduate students at Iowa State carry out various theoretical studies and observational programs, including the observation of late-type stars in micron wavelengths with a 36-inch airborne telescope at the Ames Research Center in California. There is also an antenna array associated with the Observatory which carries out observations at 26 MHz.

DRAKE UNIVERSITY MUNICIPAL OBSERVATORY. Des Moines, Iowa 50311. Located in Waveland Park. Open Fri during the academic year 8:45–10:30 p.m. before Oct 1 and after April 1; 7:45–9:30 p.m. Oct–April. Only families, or adult parties of 20, or individuals; *no children's groups.* Evenings automatically cancelled in bad weather. Telescope viewing after a short lecture. A series of free summer lectures is usually offered also, 9:00 p.m. Mon and Fri. Observing afterward, weather permitting. Write for complete schedule and description. The main instrument is the 8¼-inch refractor (ca. 1895) with optics by Brashear and a Warner and Swasey mounting; this has the interesting feature of interchangeable front lens elements providing color correction for visual work or for photography on non-color sensitized plates. Auxiliary equipment includes a Fecker double-slide plate holder for direct photography, a filar micrometer, and a polarizing solar eyepiece. There is also a 5-inch Brashear doublet camera mounted on the tube of the main telescope and a 35 mm SLR camera with adaptor. Other instruments include a small transit telescope, chronometers, chronograph, spectrum measuring microscope, Gaertner plate measuring engine, blink comparator, and a photoelectric photometer. This Observatory is a cooperative enterprise of Drake University and the City of Des Moines; Drake supplies the instruments and the city provides the building and land. It serves as a teaching facility for Drake and as a popular science center. The building includes a large lecture room, displays of astronomical transparencies and other exhibits, and a good meteorite collection.

GRINNELL COLLEGE. Grinnell, Iowa 50112. No regular public nights. A Questar and an 8-inch Celestron are available for instruction and student viewing. Grinnell owns an Alvan Clark lens and a transit telescope dating from the 1880s. There are some exhibits, including a good collection of meteorites.

HILLS OBSERVATORY. University of Iowa, Iowa City, Iowa 52240. At the corner of Dubuque and Jefferson, in the Physics Bldg. No public observation at the Observatory, but once a month telescopes atop the Physics Bldg. can be used; call Dept. of Physics to make a reservation. Some exhibits in the lobby. A 24-inch Boller and Chivens Cassegrain is used with a spectrophotometer; a 16-inch Celestron and a 5-inch refractor serve for teaching and observation. A 12-inch Cassegrain is being added for filter photometry. Iowa also has a radio facility, the Cocoa Cross radiotelescope located at Clark Lake Radio Observatory near Borrego Springs, Calif. A 2-element interferometer using Radio Shack FM antennas has also been constructed. An atlas of absolute spectrophotometry of stars has been compiled at the Observatory. As might be expected in a Department whose chairman is James Van Allen, there is much interest in space astronomy.

NORTH LIBERTY RADIO OBSERVATORY. Dept. of Physics and Astronomy, The University of Iowa, Iowa City, Iowa 52240. Located at North Liberty, Iowa. A 1.2-meter Cassegrain dish serves as a solar patrol, and dual 5-element Yagis monitor solar outbursts. A 16-element phased dipole array and a 3-element Yagi observe the sun, Jupiter, and galactic sources.

Other observational telescopes:
Beard Observatory, Central College
Carrol Community College
Des Moines Center of Science and
  Industry
Graceland College
Luther College
University of Northern Iowa

Larger planetariums:
Luther College, Decorah
Heitkamp Memorial P., Loras College,
  Dubuque
Southwest Iowa Learning Resources
  Center, Red Oak
Museum of History and Science,
  Waterloo

Kansas

UNIVERSITY OF KANSAS OBSERVATORY. Lawrence, Kansas 66045. Open each clear Fri night. The 27-inch Newtonian, 6-inch Clark refractor, and 8-inch Celestron are used for teaching and for the public. Studies of the polarization of cool stars, circumstellar envelopes, and globular clusters are carried out elsewhere by a member of the astronomy department. The department seems to be especially interested in undergraduate education and in public-service undertakings.

Other observational telescopes:
Cowley Community College
Fort Hays Kansas State College

Haskell Indian Junior College
Johnson Community College
Kansas State College of Pittsburg

Kansas State Teachers College
Ottawa University
Seward County Community College

Larger planetariums:
Kansas State Teacher's College, Emporia

Barton County Community College,
    Great Bend
Hutchinson Arts and Science
    Foundation, Hutchinson
L. Russell Kelee P., Pittsburg

Kentucky

BEREA COLLEGE OBSERVATORY. Berea, Ky. 40403. Atop the Science Bldg. in the center of the campus, 2 miles off I-75. Irregular open houses for public viewing about once a month, if weather permits. Phone: (606) 986-9341, ext. 597. The 16-inch Boller and Chivens Cassegrain with camera and photometric attachments is used mainly for teaching and student research projects. A Celestron 8-inch and an 8-inch Schmidt camera can be alternately attached piggyback to the larger instrument.

UNIVERSITY OF LOUISVILLE OBSERVATORY. Louisville, Ky. 40208. Located in Brownsboro, Ky. Public visits by arrangement. A 21-inch Newtonian/Cassegrain is equipped with photographic, photometric, and medium-resolution spectroscopic instrumentation. A Fabry-Perot interferometer will be added. On the site there is also a fully steerable 28-foot radio dish for 21-cm observations. Specialties include education, instrumentation development, spectral line shape studies for bright stars and for nebulae, and photometry. Dr. John F. Kielkopf writes: "The telescope was constructed by the Louisville Astronomical Society between 1935 and 1955 and operated on the property of Dr. Walter L. Moore from 1955 until 1970. It was known at that time as Star Lane Observatory, and was distinguished as the largest telescope solely dedicated to the use of young (ages 12–20) amateur astronomers. The telescope was donated to the University of Louisville in 1970. It has recently been refurbished and instrumented. The site of the Observatory is a 200-acre wildlife and bird refuge in a rural county near the Louisville area. Dedication of the facility is expected in late 1975 or early 1976."

EASTERN KENTUCKY UNIVERSITY OBSERVATORY. Richmond, Ky. 40475. Take EKU exit from I-75, then go 1.3 miles east. Visits, tours, and observation by appointment; also, write for semester schedule. The nineteenth-century 8-inch refractor is used for teaching and for exercises in astrometry. A possible dark-sky site is Clay's Ferry Campground on the Kentucky River, but the valley is rather deep there; Georgetown College 40 miles away has a planetarium that seats 52.

Other observational telescopes:
Kentucky Wesleyan University

Larger planetariums:
Rauch Memorial P., University of
    Louisville

Louisiana

LSU OBSERVATORY. Louisiana State University, Baton Rouge, La. 70803. "We do take special groups through on rare occasions; predominantly we are a research observatory. The facility is also used for graduate instruction." The 36-inch Cassegrain with f/7.6 and f/13.5 secondaries is used for photographic and spectroscopic work.

McNEESE STATE UNIVERSITY OBSERVATORY. Lake Charles, La. 70601. On the State University Farm at SR 14 and McNeese St. No public hours at present, but high school classes visit. The 18½-foot Ash Dome will soon have a 12-inch reflector in addition to a 6-inch refractor. Several astronomy courses are offered, and the telescopes will help with training in observational and photographic techniques.

Other observational telescopes:	Larger planetariums:
Grambling College	Louisiana Arts and Science Center,
St. Martin Parish School	Baton Rouge
Tulane University	Louisiana State University, Baton Rouge
	Louisiana Polytechnic Institute, Ruston
	Spar P., Shreveport

Maine

HEBRON ACADEMY ASTRONOMY OBSERVATORY. Hebron, Maine 01238. Rt. 119 from West Minot to Hebron Academy. Visits by appointment only. The 6-inch Clark refractor and a small reflector are used for teaching and popular astronomy. There are many good areas for camping and dark-sky observation.

UNIVERSITY OF MAINE OBSERVATORY AND PLANETARIUM. Orono, Maine 04473. The Observatory is next to the Student Union; the Planetarium is in Wingate Hall. Observatory open Sept–May 7–10 every clear night. Planetarium shows by appointment. An 8-inch Clark refractor is used for visual observation, and a Spitz Model A projector serves the planetarium.

Other observational telescopes:
Bowdoin College
Kents Hill School
Oxford Hills High School, South Paris

Maryland

U.S. NAVAL ACADEMY OBSERVATORY. Physics Dept., Annapolis, Md. 21402. Group showings as scheduled. Instrumentation includes a 16-inch Cassegrain with a Johnson-type photometer, a medium dispersion spectrograph, and several cameras. A planetarium is located in the Dept. of Navigation. The telescope is used for teaching, with courses in general astronomy and astrophysics; research is carried out in narrow- and broad-band photometry, eclipsing binaries, and model atmosphere calculations.

DAVIS PLANETARIUM. Maryland Academy of Sciences. 601 Light St., Baltimore, Md. 21230. This center is scheduled to open in spring or summer of 1976. It will include a planetarium with a 50-foot dome, seating 145, with a Minolta-Viewly Series IV projector. An observatory is planned, but details about it as well as schedules are not yet available. Exhibits will cover various scientific areas, including geology, human biology, and perception.

UNIVERSITY OF MARYLAND OBSERVATORY. College Park, Md. 20742. There is a marked turnoff from Metzerott Rd. Public nights 5th and 20th of each month, beginning 8:30 in summer and 7:30 in winter. This Observatory operates in connection with the extremely active Astronomy Program at the University, whose large faculty and 30–40 graduate students carry out research projects at Kitt Peak, Green Bank, and elsewhere. The descriptive leaflet for the Observatory cannot be improved upon: "The principal telescope (purchased with the aid of a grant for instructional scientific equipment from the National Science Foundation) is a 20″ reflector of the type known as a bent Cassegrain. It uses two curved mirrors (to form a large image in a much shorter space than would be required with one mirror) and then uses a flat mirror to reflect the light out the side for convenient viewing. Designed by the late Dr. Uco van Wijk of the Astronomy Program and built by L. C. Eichner of New Jersey, the telescope was installed and aligned in the summer of 1964. The East bay of the Observatory is used for a variety of telescopes. It was enlarged in 1968 and now holds a 10″ Schmidt-Cassegrain, a 12″ Cassegrain (both purchased with the aid of an NSF grant), and an 8″ reflector on loan from the National Aeronautics and Space Administration. The 12″ is quite similar in concept to the 20″. The 10″ uses both lenses and mirrors to form an image and is capable of photographing a larger region of the sky than is possible with the other telescopes. The 8″ is used for photographically determining precise positions of stars and for observations of comets. The Observatory is now building and acquiring a variety of sophisticated types of equipment both for training and for research using other telescopes. It is equipped with several photoelectric photometers for measuring the brightness and color of stars and other objects, an infrared photometer for measuring brightnesses in infrared radiation, a spectrograph for measuring the velocities and chemical composition of stars, a Fabry-Perot interferometer for measuring the velocity of clouds of gas, and an image tube camera to take photographs in much shorter times than would normally be the case. All of the above instruments except the image tube camera were designed and built by members of the Astronomy Program; other instruments are continuously being developed. However, in order to do much of the research, these instruments must be taken to larger telescopes in a better climate. As the optical observatory expanded, it was felt that students also needed to have firsthand exposure to radio astronomical techniques. Consequently, three small antennae, to be used as an interferometer, were set up and a second small building was constructed to house the electronics. The antennae, operating at a wavelength similar to that used for commercial television stations, are used to look at radio bursts from the sun. A larger antenna has been obtained from the National Radio Observatory and has just been set up; when the associated electronics are built it will be used to study the shape of out Galaxy. Currently the Observatory is used by many people. Approximately 3,500 students are taking our introductory astronomy course each year and, although not required as a part of the course, they are all provided an opportunity to use the Observatory and many of them do so. Several hundred students also take our introductory lab course every year; they use the Observatory in their courses. Approximately 25 undergraduate majors use Observatory facilities as part of some of their courses and approximately 50 graduate students majoring in astronomy use the Observatory to become familiar with various types of astronomical equipment. A few of these students also use the Observatory to carry out their thesis research; a few faculty members also use the Observatory for research. Most research, however, is carried out at other observatories because the lights of Washington make the sky very bright here and the prevalence of hazy and cloudy weather throughout the east coast makes observing rather difficult. The Astronomy Program also owns and operates a radio astronomy observatory at Clark Lake in the Borrego Desert of Southern California. Since acquisition, this has been supported by federal funds. The facility, acquired in 1964, is equipped with two low-frequency antenna systems, each two miles in length. The antennae (one is a dipole array and the other an array of log-periodic antennae) are used for studies of radio stars and the solar corona. Another array, more powerful and versatile, is under construction.

This telescope will be used for radioastronomical studies of the Sun, the Galaxy, pulsars, distant radio sources, and the ionized interplanetary medium."

DERWOOD OBSERVATORY. Radio Astronomy Group, Carnegie Institute of Washington, 5241 Broad Branch Rd., N.W. Washington, D.C. 20015. Located at Derwood, Md. The fully steerable paraboloid (18.3-meters) is used for hydrogen-line and continuum observations, and for detection of water in the galaxy.

WILLIAMS OBSERVATORY. Hood College, Frederick, Md. 21701. On the campus. No regularly scheduled public hours; arrangements for groups or for astronomical events. The telescope is an 8-inch Clark refractor. There is also a small telescope that was owned by the first president of Hood College, Joseph H. Apple, for whom the instrument is named. Professor Phyllida M. Willis points out that it is "unusual for a predominantly women's undergraduate college the size of Hood to have such a good facility."

RADIO OBSERVATORY OF THE GODDARD SPACE FLIGHT CENTER. NASA, Greenbelt, Md. 20771. A 5-element Yagi operates as an interferometer to monitor Jupiter.

GAITHERSBURG LATITUDE OBSERVATORY. National Geodetic Survey, National Oceanic and Atmospheric Administration, Rockville, Md. 20852. Can be seen by appointment; call (301) 926-0011. The principal instrument is a visual zenith telescope, which is used for determination of latitude. This is one of five worldwide stations that constitutes the International Polar Motion Service, which monitors the earth's polar motion.

Other observational telescopes:
Frostburg State College
Montgomery Community College
St. Mary's College of Maryland

Larger planetariums:
Washington County P., Hagerstown
Watson-King P., Towson State College, Towson

Massachusetts

BASSETT PLANETARIUM and AMHERST COLLEGE OBSERVATORY. Amherst, Mass. 01002. Morgan Hall, on the campus, for the planetarium. Shows and observing by appointment only for school groups; some special public programs. Equipment includes the Spitz A3P Planetarium projector, but far more unusual is the 18-inch Clark refractor (actually made by Carl Lundin in 1903). There is also a 7¼-inch refractor which was probably the first complete equatorial telescope built by the Clarks (1854). The price was $1,800—which probably translates to about $18,000 at 1975 costs, and which indicates the already intense American interest in new astronomical discoveries. It is also a measure of the bargains in equipment available to today's amateurs, with larger catadioptric instruments now on the market for less than a tenth the real cost in 1854.

FIVE COLLEGE RADIO ASTRONOMY OBSERVATORY. Five College Astronomy Dept., Room 127, Hasbrouck Lab, University of Massachusetts, Amherst, Mass. (Also: Amherst College, Hampshire College, Mt. Holyoke, and Smith.) This Dept. offers cooperative instructional programs in astronomy, and permits sharing of facilities, some of which are described elsewhere in this book (see Smith, Amherst, and Mt. Holy-

121

oke). The radio observatory itself is not open for public visits. One of the instruments is a 6-element array of 120-foot spherical dishes used for pulsar observations to meter wavelengths. The telescope for millimeter wavelengths is a 45-foot dish in a radome that is currently used for radio spectroscopy. The pulsar array has detected proper motion of some sources, the first success in such measurements for any radio source beyond the solar system. Also, sudden changes of period of very rapid pulsars have been observed, events that some astronomers attribute to "star-quakes,"—fracturing of the crust of a neutron star.

THORNTON OBSERVATORY. Phillips Academy, Andover, Mass. 01810. On Rt. 28. Open to groups on request; call (617) 475-3400. Also, there is an adult astronomy class in the fall. Equipment for teaching, observation, and projects in photography includes a fine old pier-mounted Brashear refractor in an electrically driven dome, plus an 8-inch Celestron, and a Questar. There are four smaller telescopes. Harold Parker State Forest in N. Reading, Mass., is a possible dark-sky site for amateurs.

SAGAMORE HILL RADIO OBSERVATORY. Air Force Cambridge Research Lab, Hanscom Field, Bedford, Mass. 01730. Located at Hamilton, Mass. Two Large dishes, 45.7-meter and 25.6-meter, observe radio stars and man-made satellites. Five radio telescopes patrol solar activity: two 8.5-meter paraboloids and three small dishes. Two other arrays also observe the sun.

CHARLES HAYDEN PLANETARIUM. Museum of Science, Science Park, Boston, Mass. 02114. Museum open Mon–Sat 10–5, Sun 11–5, Fri eve 'til 10; closed Thanksgiving, Christmas, New Year's. Museum admission: adults $2.50; ages 5–16, senior citizens, college students, $1.50; planetarium members and under age 5 free. Planetarium performances Tue–Sat at 11 and 2:45; Sun 12:15 and 2:45; Fri at 8 p.m.; one on Mon. Admission in addition to Museum admission $.50; children under five not admitted. This is one of the major planetariums in the country, equipped with a Zeiss Model VI projector. The Museum itself has extensive exhibits on natural history and physical science, including astronomy. The Museum owns a number of small telescopes (3 to 10 inches) for use by the staff and in courses. Public observing sessions are held for events of astronomical importance, such as eclipses.

DURFEE HIGH SCHOOL OBSERVATORY. Fall River, Mass. 02720. On top of the school. Viewing may be arranged by appointment through the Science Department of Durfee High School. Ordinarily the 8-inch Alvan Clark refractor in the turret-type tower is used only by the Durfee Astronomy Society. The telescope was made in 1888, and rests on a Warner and Swasey mounting.

ALICE G. WALLACE PLANETARIUM. 1000 John Fitch Hwy., Fitchburg, Mass. 01420. Rt. 2 to Rt. 13, then Rt. 2A to John Fitch Hwy. Building seating 120 is open 9–5 Mon–Fri. Star theater performances by appointment. Public shows scheduled on two weekends per month; call (617) 343-7900 for times and dates. Admission $1.50 adults, $.75 children and students. Equipped with a Spitz A4-RPY projector.

MILLSTONE HILL RADAR FACILITY. Lincoln Laboratory, Massachusetts Institute of Technology, Lexington, Mass. 02173. Located off Rt. 40, from Rt. 3. No regular public visiting. Equipment consists of an 84-foot fully steerable high-power radar dish and a 220-foot fixed (vertical) high-power radar dish. These are used for satellite tracking and for studies of the upper atmosphere. The 84-foot antenna was a pioneering development in tracking equipment, and was completed just in time to observe the first Sputnik in 1957. Within a few years it had been refined to the point that it could track a one-square-meter target at 5,000 nautical miles. Millstone has also served as an experimental facility for radar studies of meteors, the moon, Venus, and the upper atmosphere. Most recently it has been used by the U.S. Army for studies of the propagation of radar signals through the troposphere and ionosphere and by the U.S. Air Force for tracking deep-space satellites.

UNIVERSITY OF LOWELL.   Dept. of Physics, Lowell, Mass. 01854. Located in the Olney Science Center. No public hours at this time. An astronomy program is just getting under way here, with a course or two in astronomy and astrophysics and an active Astronomy Club. Over the last two years freshman honors physics classes have built, from scratch, a 12½-inch Schmidt-Cassegrain telescope, which will be housed in an observatory when the funds are available. Various demonstration materials are available for classes.

MARIA MITCHELL OBSERVATORY.   Maria Mitchell Association, P.O. Box 712, Nantucket, Mass. 02554. On Vestal St. Lectures Mon at 8:30 p.m. July–Aug at Coffin School on Winter St.; public viewing, if clear, at Loines Observatory, Milk St. extension, 9 p.m. Wed, July–Aug. Admission: adults $1, children $.50. The collection of observing instruments is almost like a museum exhibit. It includes a vintage 5-inch Alvan Clark refractor given to Maria Mitchell herself in 1859 by "The Women of America;" Miss Mitchell, the first in a long series of distinguished American women astronomers, actually picked this instrument over a 6-inch so that she might also have a position micrometer in order to put the telescope to good use. She used it also at Vassar, in preference to that college's 12-inch Fitz. The Observatory also owns two other Clark refractors: an 8-inch and a 5-inch. There is an 8-inch Celestron, too, and numerous smaller telescopes. In 1913 a 7½-inch Cooke triplet was added to permit photography. The whole observatory dates from 1902, when Miss Mitchell's relatives organized an association in her name in Nantucket. The original observatory building went up in 1908; in 1968 the Loines Observatory was added to house the 8-inch Clark. The first director, Miss Margaret Harwood, ran it from 1916 to 1957, at which time Dr. Dorrit Hoffleit of Yale took over. Many variable stars have been discovered here, and a great many young women have learned techniques of variable star work during the summer.

SMITH COLLEGE OBSERVATORY.   Northampton, Mass. 01060. McConnell Hall on the Smith campus houses a 6-inch refractor and a 3-inch Ross camera, used primarily by students from Smith and from the other colleges represented in the Five College Astronomy Department (Amherst College, Hampshire College, Mount Holyoke College, and the University of Massachusetts). Open house for the college community and the public is held once per month during the academic year. College facilities also include a 16-inch Boller and Chivens reflector, which is located in West Whately, Massachusetts, and used for photoelectric photometry and direct photography. Astronomical exhibits frequently appear in the foyer of McConnell Hall.

WILLISTON OBSERVATORY.   Mt. Holyoke College, South Hadley, Mass. 01075. The 24-inch reflector and the 8-inch refractor (one of the older Clark's finest, installed in the 1880s) are used for photometry and spectroscopy as well as in undergraduate teaching. The 24-inch, which is a Cassegrain designed and built by Group 128, is operated by the Five College Astronomy Department.

SEYMOUR PLANETARIUM OBSERVATORY. Springfield Science Museum, 236 State St., Springfield, Mass. 01103. Open Tue–Sat 1–5 and Sun 2–5. Morning school tours and lecture series. Many scientific exhibits. The planetarium uses a projector designed and built by Dr. Frank Korkosz. There is also a 20-inch reflector that is open for public viewing on Fri eves at 9:30 p.m. There is a small shop in the museum. It is one of the few locations in the Northeast where the public may observe through a telescope of this size.

BENTLEY COLLEGE OBSERVATORY. Waltham, Mass. 02154. Open to the public Thur nights, weather permitting. The 14-inch Celestron is a teaching instrument.

WHITIN OBSERVATORY. Wellesley College Campus, Wellesley, Mass. 02181. Two public nights per year, and groups accommodated by appointment. But the director, D. Scott Birney, adds: "Visitors are always welcome for an informal tour." Exhibits of photographs and some small items of astronomical interest left to the college by Sir William and Lady Huggins. Wellesley has two Alvan Clark refractors: a 12-inch and a 6-inch; these, along with the 24-inch Boller and Chivens reflector, are used in a very active undergraduate teaching program. Professor Birney writes: "A significant number of women in astronomy in the United States today received their inspiration and early training here at Wellesley." The Observatory's first director, Sarah Whiting, was a great innovator in teaching methods and some of the exercises that she invented are, in modified form, still in use. Astronomy is a field in which many women have made contributions of outstanding importance—more than in any other scientific field.

WOODSIDE PLANETARIUM AND OBSERVATORY. Raymond E. Fowler, 13 Friend Court, Wenham, Mass. 01984. Off Rt 1-A. This is a privately operated facility, which accommodates groups of up to 20 for $15 for groups of children, and $25 for adults. The program includes viewing in a small planetarium and observing with an 8-inch reflector if the sky is clear.

HAYSTACK OBSERVATORY. Northeast Radio Observatory Corporation, Westford, Mass. 01886. Off Rt. 40, about 8 miles SE of Rt. 3. Not open to the public but some tours arranged for astronomical groups. According to Haystack's own description: "The basic system includes a fully steerable 120-foot diameter radome-enclosed paraboloid antenna, a 1.2 megawatt high-voltage transmitter power supply, accurate frequency standards including a hydrogen maser, and an extensive wiring network for power and signal distribution.... Critical elements of the radio astronomy and radar instrumentation are contained in interchangeable environmental enclosures or 'boxes' designed for installation directly behind the Cassegrainian focus of the antenna." The facility is operated by NEROC, a nonprofit corporation of thirteen major educational and research institutions in the northeastern United States. This corporation took over the operation of Haystack from the M.I.T. Lincoln Laboratory in 1970. Observing time is allotted on the basis of written research proposals. Operating the big antenna in a radome eliminates wind-deformation of the collecting surface and simplifies construction of the mounting. Many different types of radio observations have been carried out, from spectroscopic mapping of gaseous nebulae to experiments in Very Long Baseline Interferometry, over a range of wavelengths from 21 cm to 7 mm.

GEORGE R. WALLACE, JR. ASTROPHYSICAL OBSERVATORY. Massachusetts Institute of Technology, Cambridge, Mass. 02139. Located near Westford, Mass. Open house once a year. There are reflectors with apertures of 16 and 24 inches, plus

Cutaway photo-drawing of Northeast Radio Observatory Corporation's Haystack Antenna inside its radome. (Photo-drawing: Graphic Arts Department, M.I.T. Lincoln Laboratory)

photometers and spectrometers. The 24-inch is computer-directed, and can be used to search for optical counterparts of radio sources under observation by nearby Haystack Observatory (q.v.). An M.I.T. news release upon successful testing of the system stated: "Development of the automated system fulfills one of the primary objectives behind the building of the Wallace Observatory in the first place. Designers envisioned it as a facility where new ideas aimed at making large telescopes costing tens of millions of dollars more efficient might be tested out on a smaller scale. In the automated system, built at a cost of $150,000, telescope drive motors are controlled by computer. The telescope position is always known to the computer. Lists of objects to be observed are maintained in the computer and are presented to the astronomer on a television-like monitor. The astronomer chooses the object he wishes to observe from a list on the TV screen by using an electronic pointer. The telescope automatically moves to the object and tracks it. Also under computer control are several instruments used to measure the light from celestial objects. The astronomer tells the computer, again by electronic pointer, how he wishes to use the instrument and what sort of measurements he wishes to make. The computer system then makes the measurement without aid from the astronomer. The incoming data are analyzed and stored by the computer and are displayed to the astronomer on the TV screen. Either the computer or the astronomer can make decisions to continue or terminate observations or move to another object. Throughout the observing sequence the controlling astronomer works at a console located in an observatory room with a controlled environment and not in the open air manually operating a telescope from an opened observatory dome. The computer system can move the telescope from star to star about twice as fast as a trained manual operator. Moreover, the same speed advantage accrues in operating instruments, since the astronomer normally needs to think of what it is he wants to do, then push appropriate buttons. Astronomers ordinarily work at night, sometimes in cold telescope domes; fatigue and discomfort result in error in telescope position or in instrument operations. The computer system, on the other hand, is very accurate in performing repetitive and complicated functions and frees the astronomer to plan his observing program in real-time according to the results he is getting at the moment of observation."

HOPKINS OBSERVATORY. Williams College, Williamstown, Mass. 01267. The old Hopkins Observatory building is on Main St. (Rt. 2), between Spring and Water Sts.; Observatory offices in Bronfman Science Center on Hoxsey St.; other telescopes atop Thompson Physical Laboratory. Planetarium shows followed by observing, weather permitting, Fri nights during the academic year and by appointment for groups; additional open nights once per semester. More extensive summer schedule. Call (413) 597-2351 for information. Instrumentation includes a 14-inch and two 8-inch reflectors, a 7⅛-inch refractor, and a 5-inch Carroll spar solar telescope. The refractor dates from 1852, and was probably the largest lens made by the Clarks up to that time; the mount, which is the same except for an electric motor, was made by Jonas Phelps, since the Clarks did not yet have much experience with such machine work. The Carroll Spar has a half-Angstrom filter for hydrogen-alpha observation, and this permits observation of prominences on the limb of the sun and of other photospheric patterns. A circular-slit automatic guiding system keeps it aligned on the sun. Students and faculty at Williams undertake a wide range of projects, from theoretical astrophysics and cosmology to solar eclipse expeditions. Projects have been carried out with the 130-foot Owens Valley radio telescope and the 210-foot dish at Parkes, Australia. The college offers an AB in astronomy and has a small AM program. This is the oldest extant astronomical observatory in the country, the original building of which is now occupied by the Mehlin Museum of Astronomy and the Milham Planetarium. Here are located exhibits of historical instruments such as a 3½-inch Troughton and Simms transit from 1834, a Molyneux and Cope regulator from 1834, and a Repsold 5-inch meridian transit from 1876. At times there are other displays in the Bronfman Science Center. The astronomy program at Williams, directed by Jay M. Pasachoff, is evidently the most extensive and stimulating in the U.S., among small colleges.

WORCESTER SCIENCE CENTER and ALDRICH OBSERVATORY. 222 Harrington Way, Worcester, Mass. 01604. Take Rt. 9 to Plantation Street, then Franklin St. East. The Observatory is located in Holden, 15 miles north of Worcester. Admission to the Science Center: $2 adult, $1 under age 17, group leaders free. Museum open Mon–Sat 10–5, Sun 12–5. Planetarium programs 1:15 Sat, Sun, holidays, and vacations. Omnisphere Programs Mon–Sat 3:30, Sat 11 a.m., Sun 2:45 and 3:30. Groups of 10 or more may reserve a program on weekdays. There are many exhibits and demonstrations of all kinds, including a Stonehenge model, the Astro Quiz, Fun with Light, and a Hall of Energy, and a shop with science kits and curios. The Aldrich Observatory is open to the public Tue (April–Nov) 8–10 p.m. The 16-inch reflector is purely for public science education. Rutland State Park is recommended as a dark-sky site for amateur astronomers.

Other observational telescopes:

Atlantic Union College
Brandeis University
Bristol Community College

Bridgewater State College
Bunker Hill Community College
Hampshire College
Lasell Junior College

Stonehill College
Taber Academy
Tufts University

Larger planetariums:
Natick High School, Natick
Somerset Public School, Somerset

Michigan

ALBION COLLEGE OBSERVATORY. Albion, Mich. 49224. One block south of Business I-94 at North Ingham St. Public nights several times during the school year; write to the Physics Dept. to be placed on mailing list or to arrange a tour of the Observatory. Instruments include: an 8-inch Alvan Clark refractor built in 1884, a Fauth sidereal clock, and a 4-inch meridian circle. These instruments are part of the original equipment of the Observatory, which was built in 1883. A recent addition is a solar prominence telescope built by Marvin Vann. The refractor has a new electric drive and the clock and chronograph have been restored to their original condition after years of neglect; ther meridian circle is being refurbished. Aside from the solar prominence telescope and the drive on the Clark refractor, the Observatory offers an excellent example of a nineteenth-century scientific facility. Some sources say that the 8-inch lens was the last made by Alvan Clark. All these instruments are used in an introductory astronomy course and by students for independent projects.

UNIVERSITY OF MICHIGAN. Ann Arbor, Mich. 48104. The University of Michigan has a very active and large Department of Astronomy with its facilities directed toward the teaching of astronomy and research. There are three visitors' nights each term, the specific nights arranged and announced to fit the University calendar. Write to place your name on their mailing list. The radio observatory is open from 2:00 to 4:30 p.m. on the third Sun of the months of June, July, Aug, and Sept. The University possesses a 10-inch Warner and Swasey refractor, a 15-inch Cassegrain reflector and four 8-inch Celestrons in the undergraduate observatory in Angell Hall. In the 1854 Observatory, now on the national register of historical buildings, is a 12-inch refractor made by Henry Fitz in 1854 and a 6-inch meridian circle made in Germany the same year. The University also owns a 37-inch Cassegrain reflector made by John Brashear. At Peach Mountain, 15 miles northwest of the campus, are two radio telescopes, 85 and 28 feet in diameter. The McMath-Hulbert Observatory 50 miles northeast of Ann Arbor has a number of solar telescopes. The primary optical research facilities are the McGraw-Hill Observatory on Kitt Peak in Arizona where there is a new 1.3-meter Cassegrain reflector and an 0.6-meter Schmidt on Cerro Tololo in Chile. The University participates in the programs at Kitt Peak National Observatory and Cerro Tololo Inter-American Observatory.

CRANBROOK INSTITUTE OF SCIENCE and HENRY S. HULBERT OBSERVATORY. 500 Lone Pine Rd., Bloomfield Hills, Mich. 48013. Take Woodward Ave. (SR M-1) to Lone Pine Rd. Go west 1 mile and turn right at street no. 500. Open weekdays 10-5, Sat 1-9, Sun 1-5; closed holidays. Museum open weekdays to groups by appointment 9-4 and Sat 9-11 a.m. Admission: adults $1.50; children and senior citizens $.75. Planetarium shows without extra charge Sat 2, 3, 4, 7, 8 p.m. and Sun 2, 3, 4 p.m.; Wed at 4. Other times for a fee. Observatory demonstrations without extra fee April–Nov dusk to 9 p.m.; later and other days by appointment for a fee. Children under five not admitted to planetarium or observatory. For information call (313) 647-0070. The principal observing instrument is a 6-inch refractor by J. W. Fecker which was built in 1927; there is also a Ross-Fecker 3-inch camera and a McMath drive. The 6-inch is used for some variable star work. The Institute is part of a complex, the Cranbrook Educational

Community, established by Mr. and Mrs. George C. Booth; other segments include an art museum and a garden. Exhibits in the Institute itself contain antique astronomical instruments, an orrery, and transparencies. An unusual educational resource for those in the Detroit area. Amateurs may, with permission, set up their own telescopes on the grounds.

MICHIGAN STATE UNIVERSITY OBSERVATORY. East Lansing, Mich. 48824. Open house every third Sat each month. MSU has a 24-inch reflector which can be used at a coudé focus at f/34.6, with a high-dispersion spectrograph. Work is under way to automate the operation of the telescope for photometric work. Astronomers in the MSU program carry out projects at major research centers and pursue statistical and theoretical studies. On the MSU campus at Shaw Lane and Science Road is the **ABRAMS PLANETARIUM,** with public shows Fri and Sat at 8 and 10 p.m. and Sun at 2:30 and 4. The exhibit hall is open during shows and also Mon–Fri 8:30 a.m.–12 and 1–4:30. Admission to show $1.25 for adults, $.50 for children (ages 5–12). Free displays include meteorites, planetarium instruments, clocks, space art, and exhibits on earth, Mars, and the moon. A sales counter offers astronomy books and souvenirs. The Planetarium owns a number of small reflectors and refractors that the public can use at 9 p.m. Fridays and Saturdays and at times of special interest. Shows are projected with the first Spitz STP ever made; the old Spitz A-1 is on display. MSU specializes in planetarium education and offers master's degree training in that field.

ROBERT T. LONGWAY PLANETARIUM. Flint Board of Education, 923 E. Kearsley St., Flint, Mich. 48502. Open to the public Sat and Sun, except holidays, 12–5, with programs every hour on the half hour. City and county students admitted free; otherwise $.25 for a child, $.50 adult. Special performances around Easter and Christmas; write for current brochure. School groups of 35 or more may arrange special shows. This is one of the largest planetariums in the country, seating 292 beneath an 80-foot dome; a half-ton Spitz Model B projector shows the stars. There are two 55-foot astronomical murals, and various lighted displays.

CALVIN COLLEGE OBSERVATORY. Grand Rapids, Mich. 49506. At the corner of East Beltline and Burton St. Public viewing every clear Thur evening. A 16-inch Celestron is equipped with a camera and spectrograph, and a UBV photometry system is under construction. There are also three 8-inch Celestrons. The principal use for these instruments is educational; students interested in astronomy take a degree in physics.

LANSING COMMUNITY COLLEGE OBSERVATORY. 419 N. Capitol Ave., Lansing, Mich. 48914. Get in touch for current information. The 10-inch Celestron is available for teaching and public observation.

OAKLAND UNIVERSITY OBSERVATORY. Rochester, Mich. 48063. Visiting by appointment. The 8-inch Newtonian-Cassegrain was built entirely by University undergraduates.

Other observational telescopes:

Almont Community College	Macomb Community College
Central Michigan University	Olivet College
Hope College	University of Michigan at Flint
Henry Ford Community College	Wayne County Community College
Lansing Community College	Warren City Schools
	Western Michigan University

Larger planetariums:
Robinson P., Adrian College, Adrian
Jesse H. Besser Museum, Alpena
Coldwater Community Schools
Grand Rapids Public Museum

Kalamazoo Public Library-Museum
Lansing Community College
Oak Park Public Schools
Chatterton Junior High School, Warren
John H. Glenn High School, Wayne

Minnesota

DARLING OBSERVATORY. University of Minnesota, Duluth, Minn. 55812. The Observatory is not normally open for visits, but the Marshal W. Alworth Planetarium gives shows without charge at 2 p.m. Sun. The main research telescope is a 16-inch Group 128 Cassegrain, which is being used for infrared studies. The Observatory has an impressive collection of historically interesting telescopes: a 6-inch Brashear refractor, made about 1880; a 9-inch Gaertner refractor (1900); and also a 12½-inch reflector from W. J. Luyten. There are other more recently purchased instruments for student use and instruction. An MS in physics is offered, with a Ph.D. in cooperation with the Minneapolis campus.

O'BRIEN OBSERVATORY. University of Minnesota, Minneapolis, Minn. 55455. Located at Marine on St. Croix. No public visiting. The 30-inch Astro Mechanics Cassegrain is used almost entirely for infrared astronomy. An interesting feature of this telescope is that by means of narrow-band filters it is possible to carry out observations by day as well as night. Also, the telescope is operated from a warm building, with filter selection and accurate pointing by remote control; in the case of a dim object an image intensifier makes identification certain. Also on the mount is an 11-inch reflector with the unusual Herschelian configuration; this was used successfully for infrared studies of Comets Kohoutek and Bradfield. Studies under way on infrared point sources may illuminate the early stages of the life of stars. Nearby O'Brien State Park is a possible remote site for amateur astronomers.

GOODSELL OBSERVATORY. Carleton College, Northfield, Minn. 55057. On campus along East 1st St. During the academic year the telescopes may be viewed between 9–4:30 on weekdays. Public nights (7–9 p.m.) are held monthly, as announced in local papers. Displays of meteorites and transparencies of astronomical objects. The Observatory owns a large 16.1-inch refractor made in 1891 by Brashear and mounted by Warner and Swasey. In 1878 Carleton bought an 8¼-inch refractor from the Clarks, and eight years later had them add a photographic correcting lens. According to the Warner book on the Clarks, an enlarging lens, made by Brashear, made possible some fine early solar photographs. At this time the refractors serve well for undergraduate instruction and are used for a certain amount of research in photoelectric photometry. At one time the Goodsell Observatory published the magazine, *Popular Astronomy.*

MACALESTER COLLEGE PLANETARIUM AND OBSERVATORY. 1600 Grand Ave., St. Paul, Minn. 55105. West on St. Clair from Snelling for one block, then turn right (north) for one long block to first building on right. Adult groups may arrange visits to the planetarium and, weather permitting, to the observatory. A heavy load of undergraduate teaching makes public accommodation difficult. The unusual situation, unusually interesting for the amateur telescope maker, at Macalester College requires some departure from this Directory's format. Since 1954 Mr. Sherman W. Schultz, Jr., has taught here in addition to pursuing his profession as an optometrist, and has made a quantity of instruments which are available to the instructional program. These include 6-inch refractors of 48- and 90-inch focal length, and reflectors of 6-, 8-, 12-, 16-, and 25-inch aperture, plus a number of cameras and a 10-inch lensless Schmidt. A solar telescope is being added. Each January, during an interim term at Macalester, Mr. Schultz offers a course in telescope making, and more than 250 instruments have been produced. This is clearly one of the most active centers for amateur telescope makers in the country. Visitors may also find exhibits of various kinds having to do with telescopes, astrophotography, etc. The Observatory itself must be a kind of museum of working optics and mountings.

WINONA STATE UNIVERSITY. Winona, Minn. 55987. No public hours. A 14-inch Celestron is used for teaching and student observing.

Other observational telescopes:
Bemidji State College
Moorehead State College
St. Mary's College
Southwest State College

Larger planetariums:
Science Museum and Planetarium,
   Minneapolis
Mayo High School, Rochester

Mississippi

KENNON OBSERVATORY. University of Mississippi, Oxford, Miss. 38677. Visits by appointment only. The 15-inch refractor is used for teaching. This instrument is an interesting example of the work of Sir Howard Grubb, of Dublin, who made many fine telescopes used in the British Isles in the last century (including those of Oxford University and the Greenwich Observatory). Grubb took over his father's optical business in 1868; later he joined with Sir Charles Parsons, the son of the Lord Rosse who owned the huge reflector, to found the firm of Sir Howard Grubb, Parsons, and Company. It was Grubb who made the mirror of the Crossley Reflector at Lick Observatory (q.v.) which is still in very active use, though in a different mounting than the one that accompanied it from England. Mississippi chose this refractor over a similar-sized one available from the Clarks. Had the Civil War not intervened with such devastating effect on so many southern educational endeavors, Mississippi would have had the largest refractor in the world; the 18½-inch Dearborn telescope in Chicago, whose tube is in the Adler Planetarium, was originally ordered by the University of Mississippi.

Other observational telescopes:
Jackson State College
Mississippi State University
Northwest Mississippi Junior College

Larger planetariums:
Jackson Public Schools

Missouri

LAWS OBSERVATORY. University of Missouri, Columbia, Mo. 65201. Physics Bldg. on College Ave. Write for a mimeo sheet of "astronomical activities." Viewing Fri, weather permitting, Sept–April 8 p.m. and May–Aug at 9 p.m. Astronomical movie Fri Sept–April 7 p.m., room 114 Physics Bldg. Instruments used mainly for teaching and viewing including three Schmidt-Cassegrains, a 16-inch and two 10-inch, all Celestrons. It should be noted that Missouri offers perhaps the friendliest and most comprehensive continuing program in public astronomical education of any state university. In addition to frequent and regular opportunities for observing throughout the year, the Physics Dept. sponsors the Mizzou Astronomy Club, which is open to anyone for a small fee. They possess an extensive film library, so that inclement weather cannot completely spoil any Friday evening. In 1975 the Coordinator is Dr. Terry W. Edwards, 420 Physics, phone (314) 882-3036, who advises: "Please feel free to ask questions at any time." In the late nineteenth and early twentieth centuries the Observatory owned several refractors by various makers. Rock Bridge State Park is mentioned as a place with darker skies that amateurs might use.

MORRISON OBSERVATORY. Central Methodist College, 700 Park Rd., Fayette, Mo. 65248. Open to the public Thur evenings (when clear) during the school year. Small exhibit of star globes and charts. The 12-inch Clark refractor and 12-inch reflector are used for observation, labs, and training of astronomy majors. There is also a 6-inch meridian circle on exhibit, though not in use now. The refractor is one of the larger fine instruments of this type available for public observation.

R. A. LONG PLANETARIUM. Museum of History and Science, 3218 Gladstone Blvd., Kansas City, Mo. 64123. Shows Sat and Sun 2, 3, and 4 p.m.; school groups at 9:30, 11:00, 1:15. Admission: $.80 per child for classes; weekend shows $.75. Murals and photographs of astronomical objects. For dark-sky sites and possible telescope-viewing opportunities, get in touch with the Kansas City Astronomical Society at the Observatory address.

WASHINGTON UNIVERSITY OBSERVATORY. St. Louis, Mo. 63130. The Physics Dept. suggests, quite reasonably, that in view of demand by students on this small observatory, most would-be observers would do better to go to McDonnell Planetarium in Forest Park, which has larger telescopes for public use. Washington University does own a 6-inch Alvan Clark refractor and a 7-inch Questar, which are used for teaching. A refractor freak could probably arrange to see the 6-inch. "There is a strong interest in astrophysics at the research and graduate level" and also undergraduate astronomy courses.

McDONNELL PLANETARIUM. City of St. Louis, 5100 Clayton Rd., St. Louis, Mo. 63110. Located in Forest Park. Write for current information and schedules for all the astronomical activities of this excellent center. Telescopic observation by the general public Tue at 9:15. Equipment for this includes three 8-inch Tinsley Cassegrains and one 12½-inch home-built telescope. In the past there have also been viewing sessions Fri and Sat evenings. A short introduction in the Star Chamber describes how a telescope works and what will be seen. Admission is $.50 per adult and $.25 per child. Except for Thanksgiving, Christmas, and New Years, there are two to five planetarium shows daily, including night shows Tue, Fri, and Sat. Phone (314) 535-5810 for more information. There are courses in a variety of subjects having to do with astronomy, at all levels; write for a course brochure. The St. Louis Astronomical Society, Inc. invites members of all ages, who pay dues and attend monthly meetings at the Planetarium. There are numerous and constantly changing exhibits and films. A bookstore and gift shop offers a large selection of material about astronomy. This center ranks with Adler, Chabot, Griffith, Morehead, Fernbank, and several in the Northeast as among the best.

Other observational telescopes:
Rockhurst College
Southeast Missouri State College
Southwest Missouri State College

Larger planetariums:
Tarbell P., Inca Cave Park, Laquay
Kansas City Museum of History and Science

Montana

MONTANA STATE UNIVERSITY. Physics Dept., Bozeman, Mont. 59715. On top of the Physics Bldg. No regular public hours, but the public is invited during events of special interest. Equipment, consisting of an 8-inch Celestron, a 10-inch Cave reflector, and a 3-inch refractor, is used for teaching; some photographic and spectroscopic work has been done. Gallatin National Forest to the south is full of dark-sky campsites.

BLUE MOUNTAIN OBSERVATORY. Dept. of Physics and Astronomy, University of Montana, Missoula, Mont. 59801. Located on the peak of Blue Mountain, SW of Missoula. Not open to the general public. A 16-inch Boller and Chivens f/18 Cassegrain is used with a Johnson single-channel photoelectric photometer for variable star studies. Also, the observatory is used for instructional purposes.

Other observational telescopes:
Carroll College Observatory

Nebraska

BOSWELL OBSERVATORY. Doane College, Crete, Neb. 68333. On the campus. Visits on request. This Observatory, built in 1883, owns a fine 8-inch Clark refractor, and also a transit instrument. With three or more campgrounds in the area a side trip here might be possible for a continent-crosser on I-80.

BEHLEN OBSERVATORY. Dept. of Physics and Astronomy, University of Nebraska, Lincoln, Neb. 68508. Located at the University of Nebraska Field Laboratory near Mead, Neb., approximately 1.5 miles north of SR 63 and 9 miles east of US 77. Public nights currently once in the fall and once in spring; prior to loss of University funding, public nights were every other Friday, with slide lectures and viewing through a 30-inch telescope. The main instrument is a Boller and Chivens 30-inch Cassegrain with a spectrograph and a camera built by the same company. A photoelectric photometer and a spectrum scanner have been built in their own shops. An area scanner and a two-star photometer are planned. Special studies are carried out in variable stars, binaries, interstellar matter, clusters, and instrument development. A correspondent writes: "In this age when most research telescopes are paid for with government funds, our 30-inch is almost unique in that it was the gift of Walter D. Behlen, a philanthropist from Columbus, Nebraska.... Many exhibits at the Morrill Hall Museum were also the gift of Mr. Behlen."

NEBRASKA WESLEYAN OBSERVATORY. Nebraska Wesleyan University, Lincoln, Neb. 68504. At 50th and St. Paul in Lincoln. Public visits may be arranged. The 10-inch Celestron and 8-inch Cave reflector are used for teaching.

CREIGHTON OBSERVATORY. Creighton University 2500 California St., Omaha, Neb. 68178. Groups of ten to twenty may arrange to use the eight Criterion 6-inch reflectors which can be set up on the open deck on top of the new Rigge Science Hall. Preference is given to high-school and adult groups interested in simple, beginning observations. Groups of ten or fewer may arrange for more advanced observation in the older (1885) Observatory, which is used for instructional purposes. The Observatory has a 5-inch classical refractor of extralong focal ratio of 17.5 for lunar and planetary observation, which occupies a 15-foot dome. A Fauth 3½-inch transit scope is housed in an adjacent part of the building. On the Observatory balcony two 6-inch reflectors and a 10-inch Celestron are set up for special purposes. The Observatory owns several instruments of historical interest, including pendulum solar and sidereal clocks, a chronograph, a large celestial globe, a sextant, an early Steinheil diffraction grating spectroscope and a Stewart micrometer. The collection of instruments illustrates some of the history of astronomy.

Other observational telescopes:
Hastings College
Osaacs Science Center
Union College

Larger planetariums:
Lueninghoener P., Midland Lutheran
College, Fremont

W. J. Arrasmith P., Grand Isle High School
Jensen P., Nebraska Wesleyan University, Lincoln
University of Nebraska Museum, Lincoln

Nevada

Observational telescopes:
Clark County Community College
Nevada Southern University
University of Nevada at Las Vegas

Larger planetariums:
Fleischman Atmospherium, University of Nevada, Reno

New Hampshire

SHATTUCK OBSERVATORY. Dartmouth College, Hanover, New Hampshire. Information not supplied about current instrumentation, visitors' programs, and teaching. D. J. Warner's book on the Clarks gives much historical information, including the fact that a spectroscope actually used for almost a hundred years is on display in the Dartmouth College Museum; this was originally made by the Clarks. A second spectroscope is also on display, the one with which the sun's reversing layer was discovered by Charles A. Young, on an eclipse expedition to Spain in 1870. Dartmouth also acquired a 9.4-inch Clark refractor.

Other observational telescopes:
University of New Hampshire

New Jersey

MORRIS MUSEUM ASTRONOMICAL SOCIETY. P. O. Box 125, Convent, N.J. 07950. The Museum is open 10–5 weekdays and 2–5 Sun, but closed Sun and Mon during July and Aug. Use of telescopes is by members of the Society, who pay dues of $12 per family per year. Workshops for all age groups, and for proficiency levels from nontechnical to highly specialized. The first Sat in Nov is Astronomy Day, for the public. The telescopes are used on a platform on the roof. They consist of an 8-inch Celestron and two small instruments. Solar and radio telescopes are under construction. Persons who join the Society are kept informed of observing sessions and projects by a newsletter.

CRAWFORD HILL OBSERVATORY. Bell Telephone Laboratories, Holmdel, N.J. 07733. On Crawford's Corner Rd. Not open to the public. Radio telescopes consist of a 20-foot horn reflector antenna and a 7-meter offset Cassegrain millimeter wave antenna. At this time the Observatory specializes in extremely detailed observations of organic molecules in the interstellar medium, such as cyanide and carbon monoxide. Radio astronomy was born at this laboratory (see Green Bank) when Karl Jansky, attempting to account for some disturbances to telephone communications, discovered radio waves coming from the center of the Milky Way Galaxy. Later, the isotropic background radiation permeating the Universe was discovered, lending support to the Big Bang theory and leading Fred Hoyle to reconsider the Steady State theory.

DREW UNIVERSITY OBSERVATORY. Madison, N.J. 07940. On Rt. 24 in Madison, on top of the Hall of Sciences. Open to the public Fri during the school year, 7:30–10:30 p.m. and at other times coinciding with special celestial events. A 10-inch Celestron and four Questars are all used mainly for visual observing by students and public, although there are also cameras and photoelectric equipment for undergraduate research programs. The main telescope is housed in a 16-foot dome located on a 40-foot-square observing deck. A radio telescope for observation of the sun at 10 cm is under construction, and there are hopes for a 12-inch optical instrument for research.

NEWARK MUSEUM PLANETARIUM. 43–49 Washington St., Newark, N.J. 07101. Open 12–5 daily except holidays. Shows at 2 and 3 p.m. Sat, Sun, and Holidays, Sept–June. Shows at 12:15 Mon and Wed, July–Aug. School groups of 15 or more can arrange shows between 10 and 2 p.m. Mon–Fri. Admission for shows: adult $.50, children $.25. Children under 7 may not attend shows. Exhibits include historical displays about the great astronomers and a lunar landscape; there is also a heliostat. The small observatory with its 6-inch catadioptric telescope is not at this time used for public observing sessions. There is, however, a program of Sat morning and summer courses for high school and junior high students which includes nighttime observation with this instrument. Write for further information.

STOCKTON COLLEGE OBSERVATORY. Natural Science Division, Stockton State College, Pomona, N.J. 08240. On Atlantic County Rt. 575, 1 mile NE of Pomona. Usually open to visitors on clear Fri evenings, but check beforehand, calling (609) 652-1776. There are 14- and 8-inch Celestrons, plus some smaller Edmund reflectors, and some 35 mm cameras. These are used for astronomy courses and also for amateur astrophotography.

PRINCETON UNIVERSITY OBSERVATORY. Princeton, N.J. 08540. Ivy Lane, off Washington Rd. from Route 1; in Peyton Hall. Write for current open-house announcement. Six evenings per year, 7:30–9:30 p.m. Lectures at 8 p.m. and repeated if warranted. Special arrangements for groups; write, giving size and age range of group, to Administrative Officer. A 4½-inch and a 9-inch refractor made by Alvan Clark are used for teaching and for guest observations. A 36-inch Boller and Chivens reflector serves for advanced teaching and research. Princeton specializes in theoretical astrophysics and space astronomy. On display is the Rittenhouse Orrery, and on weekdays the University bookstore is open 9–5:30.

SETON HALL UNIVERSITY. South Orange, N.J. 07079. On top of McNultey Hall on campus. Open Tue and Wed 9–11 p.m. A 16-inch reflector, an 8-inch Celestron, and a 5-inch Schmidt camera are used for student observing and photography.

Other observational telescopes:
Bergen Community College
Jersey City State College
Rider College

Larger planetariums:
The Peddie School, Hightstown
New Jersey State Museum, Trenton

New Mexico

CAPILLA PEAK and CAMPUS OBSERVATORIES. University of New Mexico, Department of Physics and Astronomy, Alberquerque, N.M. 87131. Capilla Peak is in the Manzano Mountains, at 9,390 feet elevation, 60 miles south of the campus. Occasional special tours are arranged for recognized amateur groups. At Capilla Peak there is a 24-inch Boller and Chivens Cassegrain with a digital-counting photometer that measures starlight photon by photon. The telescope is used to study light pulsations of the Crab Nebula Pulsar and the light curves of X ray emitting binaries; also, there is equipment for image-tube spectrography and for solar spectroscopy. On the campus, there is a 15-inch reflector that is open on clear Thursdays at 8 p.m. during the academic year.

NEW MEXICO STATE UNIVERSITY. Dept. of Astronomy, P.O. Box 4500, Las Cruces, N.M. 88003. On the campus the **CLYDE W. TOMBAUGH OBSERVATORY** holds public nights scheduled each year; write for exact dates and times. The 5¼-inch refractor and two 12-inch Cassegrains (one of which is the property of the Las Cruces Astronomical Society) provide training for undergraduates and viewing for local school groups. Write Astronomy Dept. for group arrangements and about Society membership. Anyone driving through Las Cruces will notice the dome of the **TORTUGAS MOUNTAIN OBSERVATORY** atop the hill just east of town, the summit of which is decorated with a huge collegiate "A" in white. Tortugas Mountain itself looks oddly small against the spectacular jagged backdrop of the much higher peaks of the Organ Mountains behind it. Yet it is nearly 5,000 feet up. The Boller and Chivens 24-inch and a pair of smaller reflectors (16- and 12-inch) are mostly used for planetary studies and have accumulated large collections of photographs of the planets. Steady air and the very long focal length of the 24-inch (150-feet, f/75) have produced some fine studies of such things as Jupiter's Red Spot, Mercury's surface features (large scale, necessarily), and the structure of Venus's atmosphere. This Observatory and the other research area at **BLUE MESA** are only open by special arrangement. At Blue Mesa Observatory, 6,600 feet above sea level and 30 miles NW of Las Cruces, a 24-inch Cassegrain of shorter focal length (f/15) carries out programs in stellar photoelectric photometry and spectroscopy, as well as being used to make photographs. It is especially good for recording star clusters on a large scale while preserving image sharpness; its fused quartz mirror and very solid English mounting are responsible for this. Much sophisticated auxiliary equipment makes possible ultraprecise spectrographic work. In this very active program there are numerous undergraduates, many graduate students, and a broad array of research programs carried out by the faculty. Nearby, an extensive daytime visit is possible at the Sacramento Peak Observatory (q.v.). On the first Sunday in October there is an annual tour of Trinity Site at Alamogordo, the crater left by the first atomic explosion. The best nearby camping areas are probably those some distance west, in the Gila National Forest (headquarters in Silver City, N.M.).

CORRALITOS OBSERVATORY. Northwestern University, Dept. of Astronomy, Evanston, Ill. 60201. Located 20 miles west of Las Cruces, N.M. The largest instrument here is a 24-inch Cassegrain; it can be pointed automatically by a computer and can be used with an image orthicon and a video storage system. A UBV photometer can also be attached. Another remote-control telescope is a 16-inch image-orthicon-equipped Cassegrain used for closed-circuit TV displays. On a single mount is a 12-inch Cassegrain and a 6-inch f/4 Schmidt with all-reflecting optics; the Schmidt is one of very few in the world of such unusual configuration. A recent program on the 24-inch was a series of experiments with image orthicon photometry; previously, the telescope has been used for supernova searches, comparing "live" pictures with a library of photographs made on the same instrument. The method has turned up quite a number of supernovae. A program to look for transient phenomena on the surface of the moon has been completed; some years ago there had been reports of gaseous emissions—reddish clouds— on the moon's surface.

NASA–LANGLEY RESEARCH CENTER METEOR OBSERVATORY. Hampton, Va. 23365. Located 12 miles east of Las Cruces, N.M. off US 70 at Organ Pass. During the years 1968–72, various kinds of meteor patrol equipment were set up here. These included 21 slitless Maksutov spectrographs and K-24 aerial cameras equipped with chopping shutters, time displays, and programmers. This site, a mile above sea level, has very clear skies; it previously was used for a super-Schmidt meteor patrol and later for one of the Smithsonian's Baker-Nunn satellite tracking and laser ranging stations. Originally the purpose of the patrol was to try to estimate the likelihood of a manmade spacecraft's encountering a meteoroid. Later, attention turned to the problem of determining the constituents of meteoroids, and 764 spectra of meteors helped identify the elements present.

LANGMUIR LABORATORY FOR ATMOSPHERIC RESEARCH and JOINT OBSERVATORY FOR COMETARY RESEARCH. New Mexico Institute of Mining & Technology and NASA-Goddard Space Flight Center, Socorro, N.M. 87801. Visits to Langmuir Laboratory can be arranged by telephoning (505) 835-5423 in advance during the period June 15 to Aug 31. No visitors' program available at comet lab or telescope. All of these facilities are located near 10,600-foot South Baldy Peak in the Magdalena Mountains about 17 air miles west of Socorro. Access from Socorro is gained by traveling 16 miles west on U.S. 60, turning left to the Water Canyon Campground, and left again for approximately 20 miles of rough, unpaved road. A pickup or four-wheel drive vehicle is recommended for the last 20 miles. The comet observatory houses a 14-inch, F-2, Schmidt camera to photograph comet tails and a general purpose 16-inch Cassegrainian-Newtonian. There is also a fully automated telescope under construction on the mountain. This facility is linked by microwave to the campus. A computer directs the telescope to scan galaxies to pick up supernova outbursts. When the telescope is operating at its designed rate of 1,000 galaxies per hour, it will present an eerie spectacle to the observer as it swings every few seconds to settle on a new object.

VERY LARGE ARRAY. National Radio Astronomy Observatory, Green Bank, W.Va. 24944. Located on the Plains of San Augustin on US 60, 40 miles west of Socorro, New Mexico. When this project is completed in 1981, it will be the largest single-site radio observatory in the world. Standard-gauge railroad tracks are being laid out in an enormous Y-shaped pattern, each arm of which is 21 km (13 miles) long. Altogether there will be twenty-seven solid-surface radio reflectors, each 25 meters (82 feet) in diameter. These will be distributed along the tracks at observing stations, sets of rails perpendicular to the transport rail. By placing them at different stations and connecting them in different ways, radio astronomers will be able to synthesize all manner of apertures. Maximum

Artist's conception of the Very Large Array on the plains of San Augustin, New Mexico. Looking east, parallel to U.S. 60, one arm of the Y-shaped array stretches 13 miles; near the assembly building, one antenna is being carried to another observing station. (National Radio Astronomy Observatory drawing)

effective aperture will be 17 miles. As a radio source crosses the sky, millions of measurements of it, using different combinations of antennas, will build up a detailed picture. The Plains of San Augustin are themselves a remarkable spectacle. In February, 1975, their arid vastness made initial structural efforts on the VLA look Lilliputian; they stretch 20 and 30 miles in the middle of a ring of low mountains, high and dry at 7,000 feet. There is almost no traffic on US 60 and almost no human activity other than a little ranching. Astronomical travelers should take this route until the Observatory is in operation, at which time automobile ignitions will be a source of interference. This is, perhaps, the most beautiful remaining "wide open space" in the country.

Larger planetariums:
R. H. Goddard P., Roswell Museum,
  Roswell

New York

DUDLEY OBSERVATORY. 100 Fuller Rd., Albany, N.Y. 12205. Near the SUNY at Albany campus, at the corner of Fuller and Railroad Aves.; take throughway exit 24, Northway Exit from I-90, or Rt. 5E; the Observatory is between Rts. 2 and 20 on Fuller. Visits by prior arrangement only. For a membership fee, one may join the Friends of the Observatory, which meets for lectures. A local astronomy group also meets at the Observatory on the second Mon each month. Other group lectures may be arranged for a fee. The Library maintains exhibits throughout the building, and has a bookstore. The Observatory owns a Clark "comet-seeker" and a 12-inch Brashear refractor. In 1973, a

16-inch Boller and Chivens Cassegrain was installed on the roof of the Earth Science Building. This is equipped with a photoelectric photometer and a spectrograph. Mrs. Christine Bain, librarian, writes; "Optical work is now done chiefly at the national observatories." Near Bolton Landing, N.Y., a 100-foot alt-azimuth radio telescope is being completed and tested. It will be used to wavelengths as short as 6 cm. A variety of research programs are under way at the Observatory, by members of the SUNYA astronomy department, or by the Space Astronomy Laboratory. One particular specialty is the study of "space dust," collected by balloons, satellites, rockets, and space-craft. A piece of apparatus called an "ultrasensitive oscillating fiber microbalance" is used to measure growth and sublimation rates of ice particles in space.

ALFRED UNIVERSITY OBSERVATORY.   Alfred, N.Y. 14802. Public hours 9–11 p.m. on clear Fri, Sept–May. Tours can be arranged. On the New York campus are a 16-inch f/11 Ealing "Educator;" a 9-inch refractor, whose objective is perhaps the last one made (about 1863) by Henry Fitz, in a modern mount with a variable-frequency drive; several 6-inch reflectors; a 20-inch f/5.5 reflector; a 14-inch f/6 reflector under construction; some 4-inch refractors; and an f/11 photometer. A spectrograph and heliostat are under construction. Two 12-inch reflectors of long and short focus are designated for a site on San Salvador Island in the Bahamas, where there is a midwinter program. All these instruments are at the service of undergraduates, whether they are science majors or not; "we are quite unusual (perhaps unique) in the amount of equip-ment we make available for hands-on use." Photographs of the campus observatory domes look like those of a miniature Kitt Peak. This is one of the best teaching facilities in the country.

KELLOGG OBSERVATORY.   Buffalo Museum of Science, Humboldt Park, Buffalo, N.Y. 14211. Observatory open Fri nights Sept–April; solar division open daily Mon–Fri during July and Aug. Special day and night tours by arrangement. An 8-inch f/15 re-fractor figured by Carl Lundin and a 6-inch f/10 Gee refractor serve amply for public observation. A polar heliostat with a 7¼-inch f/26 refracting lens provides research capability in solar work; a 4.5-Angstrom bandpass filter permits detailed hydrogen-alpha observation, and there is a 6-foot spectrograph/spectroheliograph. Some lunar and planetary studies are carried out, but mostly this is a teaching and public service observatory. The Museum features a Hall of Astronomy, other exhibits on astronomy in a planning stage, and a bookstore. There are no fees, and this is an unusually fine public center for astronomical education.

VANDERBILT PLANETARIUM.   180 Little Neck Rd., Centerport, Long Island, N.Y. 11721. Open Tue–Sun and Monday holidays; with telescopic observation every evening after planetarium show. Schedule of lectures varies with season. The 16-inch Cassegrain and the 4½-inch Mogy refractor are used primarily for public observation. Facilities include a very large planetarium theater, a 4,000-square-foot exhibit hall, and an astronomical library. Vanderbilt is a major institution for public education in astronomy. A nearby campground is Sunken Meadow State Park.

FUERTES OBSERVATORY.   Cornell University, Ithaca, N.Y. 14850. No set times for any public hours. There is a 12-inch refractor. Work in theoretical astrophysics and radio astronomy goes on at Cornell. More information was available about Cornell's association with the Arecibo Radio Observatory (see the entry under Puerto Rico).

CORNELL RADIO ASTRONOMY OBSERVATORY. Space Science Bldg., Ithaca, N.Y. 14850. Located at Danby, N.Y. A 26-meter spherical dish is used for general radio mapping of the sky; a 5.2-meter reflector serves for flux measurements and testing.

HARTUNG BOOTHROYD OBSERVATORY. Space Science Center, Cornell University, Ithaca, N.Y. 14853. Five miles east of the campus on Mt. Pleasant Rd. No public nights. The 25-inch Cassegrain is used for infrared astronomy and for a visual occultation program.

MARIETTA OBSERVATORY. State University of New York at Morrisville, N.Y. 13408. Scheduled class observations and public or club sessions by arrangement with the Mathematics and Science Division. The 12-inch Newtonian and 8-inch Schmidt-Cassegrain are used for instruction and observing. A very unusual feature of this college is that the library has five loan telescopes of the Maksutov 3-inch type; Project L.O.O.K. (Library Oriented Observation Kits) pamphlets are provided to aid the novice. The 12-inch reflector is housed in an actual silo that was donated by the Madison Silo Co. (see February 1975 *Sky & Telescope*).

COLUMBIA UNIVERSITY. New York, N.Y. 10027. Information not supplied on Columbia's observatory at Harriman, New York. That observatory has a 24-inch telescope, presumably a reflector. Astronomers associated with Columbia run projects at Kitt Peak, Cerro Tololo, Green Bank, etc.

RUTHERFURD OBSERVATORY. Columbia University, New York, N.Y. 10027. Located in Pupin Hall, 538 W. 120th St. Public hours first Fri in each month, Sept–June, nightfall to about 9:30. The refractor, which is used for teaching and public observation, may contain an 11½-inch lens with which Lewis Morris Rutherfurd used to photograph the stars from New York City in the late 1850s. Columbia describes this instrument as a "12-inch Clark refractor." In any event, it is probably the best kind of instrument for viewing the moon and planets from the middle of Manhattan.

HAYDEN PLANETARIUM. 81st St. at Central Park West, New York, N.Y. 10024. Take 6th or 8th Ave. Independent Local subway train to 81st St. station. West-side bus lines to 81st St. Some automobile parking. Shows all year Mon–Fri at 2 and 3:30 p.m. and Sat–Sun at 1, 2, 3, and 4 p.m. Additional shows depending on season. This famous institution includes all manner of exhibits, lectures, and academic courses. Write for complete current schedules and course offerings. In light of New York City's sky conditions, the Hayden Planetarium offers the most feasible approach to astronomy for city-bound residents. Reproductions and recordings bring culture to the provinces; a fine planetarium brings dark skies indoors. Columbia University's Rutherfurd Observatory is almost the only place inside the city for any kind of direct visual observation. For field trips to darker skies, see entries in Connecticut, New York, New Jersey, and Pennsylvania. See also Vanderbilt Planetarium (N.Y.).

VASSAR COLLEGE OBSERVATORY. Poughkeepsie, N.Y. 12601. Located on Raymond Ave. in Poughkeepsie. Visits by appointment. There is a 15-inch reflector used for photometry and spectroscopy, and an 8-inch refractor for visual observation. According to Deborah Jean Warner, a 12½-inch refractor made by Henry Fitz was completely worked over by Alvan Clark and Sons; they reground the lens, in effect using Fitz's work as optical blanks, and repaired the clock drive. Later it was entirely remounted by Warner and Swasey. Finally, Vassar gave this instrument to the Smithsonian. Warner also mentions that Vassar owned telescopes of smaller aperture: a meridian instrument by Young of Philadelphia, with two Clark collimating telescopes; and Clark portable refractors of 3- and 6-inch apertures. Vassar is rich in astronomical history. One might mention that the great days of the Observatory coincided with the 24-year tenure as director of Maria Mitchell (see: Maria Mitchell Observatory, Massachusetts). Vassar himself brought her there after she had distinguished herself by winning the King of Denmark's gold medal for the discovery of a comet invisible to the naked eye; her modesty almost caused her not to register a claim. Her own telescope (a 5-inch Clark) was among those used by the students.

C. E. KENNETH MEES OBSERVATORY. University of Rochester, Rochester, N.Y. 14627. Call (716) 275-4385 for a Sat appointment, May through Oct, only. The 24-inch Boller and Chivens Cassegrain, with its auxiliary equipment is used for programs in rapid photometry, infrared photometry, and electrophotography. The Observatory is primarily a research facility but is also used as a teaching tool. Ontario County Park Campgrounds offer a possible location for amateur observation.

SYRACUSE ASTRONOMICAL SOCIETY OBSERVATORY. 1115 E. Colvin St., Syracuse, N.Y. 13210. The Observatory is located off Strong Rd., about 2 miles south of Vesper, N.Y. Take Rt. 81 to intersection with Rt. 80, then Rt. 80 west to Strong Rd., then south to Observatory Rd. About two public nights per month, April–Oct, and by request; write for schedule, and also to arrange for projects. A 16-inch reflector is housed in an arched roll-off roof. There are four level concrete pads on which other instruments may be used.

SYRACUSE UNIVERSITY OBSERVATORY. 201 Physics Bldg., Syracuse, N.Y. 13210. On University Place. Public viewing Mon 7:30–10 p.m. when the sky is clear. The Observatory has a fine 8-inch Clark refractor that has been there since its dedication in 1887. An astronomy text is available in the campus bookstore.

RENSSELAER OBSERVATORY. Rensselaer Polytechnic Institute, Troy, N.Y. 12181. On 15th St. in Troy. The 12-inch Newtonian, 6-inch reflector, and Ross camera are all used exclusively for teaching. BS, MS. and Ph.D. offered in physics.

Other observational telescopes:
Bronx Community College
Broome Community College
C. W. Post College
College of St. Rose
Concordia Collegiate Institute
Dowling College
Hartwick College
La Guardia Community College
Nassau Community College
New York University

Queensborough Community College
Skidmore College
SUNY at Buffalo
SUNY at Stonybrook
Suffolk County Community College
Utica College

Larger planetariums:
Henry Hudson P., SUNY at Albany
Roberson Center for the Arts and
    Sciences, Binghamton

Newburgh Free Academy, Newburgh
McGraw-Hill P., 330 W. 42nd St., NYC
Grymes Hall, Wagner College, Staten
   Island
Strasenburgh P., Rochester

Schenectady Museum
North Rockland High School, Thiells
West Islip High School
Hudson River Museum, Yonkers

North Carolina

MOREHEAD PLANETARIUM of the Morehead Foundation and its OBSERVA-TORY, operated by the Dept. of Physics and Astronomy. 278 Phillips Hall. University of North Carolina, Chapel Hill, N.C. 27514. Located on UNC Campus, Franklin St. This planetarium, with a seating capacity of 470 and its Zeiss Model VI Sky Projector, is one of the largest in the country. Shows are at 8 p.m. Mon–Fri with additional shows June–Aug at 11 a.m. and 3 p.m. Also: Sat at 11 a.m. and 1, 3, and 8 p.m.; Sun 2, 3 and 8 p.m. Home football Sat at 11, 5 and 8. Admission: adult $1.25; student $1; children $.75. Exhibits in the hallways around and beneath the planetarium are free of charge and are open daily 2–5 and 7:30–10 (Sat opens at 10 a.m. and Sun at 1 p.m.). The building in which the Planetarium and Observatory are housed, with the rose garden and giant sundial in front of it, adds one more touch of neoclassical graciousness to this oldest State University Campus. The Pantheon-like dome over the Planetarium makes Roman architecture serve modern science. The exhibits on astronomy and mathematics are well presented, and the Planet Room (a Copernican Orrery, with planets modeled to relative scale moving in orbits around the ceiling of the dimly-illuminated room) may be one of the best exhibits of its kind anywhere. The wood-paneled Rotunda Art Gallery with its fine portraits and elaborate striking clock recall the appointments for some earlier observatories, patronized by royalty, in Europe. Next to the Morehead Building are botanical gardens and the whole UNC campus. This is worth an extensive detour for a visit; it is one of the pioneering institutions in popular astronomy, established as such on the advice of Harlow Shapley, and also of the astronomer E. E. Barnard who recognized the impracticality of a great observatory in the muggy Southeast. The Observatory is located in a wing of the same building and is open for visitors' nights about every other Fri Sept–April, 8–10 p.m. Since only 35 can be admitted, though, free tickets must be secured in advance by writing after Aug 1 for Sept–Nov; after Nov 1 for Dec–Feb; and after Feb 1 for March–April. Send stamped self-addressed envelope (or write for schedule) to: Guest Night, University of North Carolina, Dept. of Physics and Astronomy at address above. Special tours by groups interested in astronomy may be arranged on other than Guest Nights by writing to the same address, and daytime looks at the telescope may also be arranged. The main telescope is a Boller and Chivens 24-inch Cassegrain which is used for photoelectric photometry of variable stars and galaxies, and photographic photometry of quasars. The Observatory possesses several instruments of more than usual historical interest, since their presence establishes UNC's claim to the first astronomical observatory in the United States. The first president of the new University, Joseph Caldwell, spent about half of his entire budget for equipment on astronomical instruments in the course of a trip to Europe in 1824. These included a 3-inch transit and a 2½-inch altazimuth refractor made by Simms of London; and another small telescope made by one of the Dollonds. Together with the armillary sphere and sidereal clock purchased at the same time, they provide a rare collection of what was available towards the end of the eighteenth century in the way of smaller instruments and devices demonstrating astronomical craftsmanship. An amusing anecdote, recorded by

G. Edward Pendray in *Men, Mirrors, and Stars*, recounts the efforts of some Chapel Hill faculty to hide valuables in the tubes of these instruments (by that time disused and in storage) to preserve them from Yankee soldiers who occupied the campus at the end of the Civil War. It did not work: the soldiers found and made off with the watches and jewelry anyway. Fortunately the telescopes themselves looked worthless to the intruders! In the end, the soldiers' commanding officer made them return the valuables.

DUKE UNIVERSITY. Physics Dept., Durham, N.C. 27706. A roll-off roof on the top of the Physics Dept. houses a number of telescopes that are used almost exclusively for teaching purposes: five 8-inch reflectors and a 12½-inch reflector. The Department also owns three Celestrons, a 6-inch refractor, and various accessories. There is a small phase-switched radio interferometer; the first millimeter wavelength radiation from the sun was observed at Duke.

UNIVERSITY OF NORTH CAROLINA AT GREENSBORO. Greensboro, N.C. 27412. Located on top of the Graham Bldg. on Spring Garden St. Open Thur evenings during the academic year, after nightfall, when clear. The 10-inch Celestron is mainly used for teaching and for public observing, but is equipped with a 4-inch Celestron guide-scope and other accessories for photography. There is also a portable 3-inch refractor with a key-wound clock drive. The Physics Dept. also has a 12-inch Newtonian in a roll-off roof observatory on top of Petty Building; experiments in photography and spectroscopy are planned for this instrument, and it will be used for student research projects. The Greensboro Astronomy Club meets regularly in the Graham Bldg. to hear guest speakers.

Other observational telescopes:
Belmont Abbey
Davidson College
Elon College
Guilford College
Lenoir Community College
University of North Carolina: Asheville,
  Charlotte, East Carolina, North
  Carolina State

Larger planetariums:
Children's Nature Museum, Charlotte
Central Carrabus High School, Concord
Schiele Museum of Natural History,
  Gastonia
Greensboro Natural Science Center
Robeson County Board of Education,
  Lumberton
Margaret C. Woodson P., Salisbury

North Dakota

UNIVERSITY OF NORTH DAKOTA. Department of Geography. Grand Forks, N.D. 58202. A 16-inch Cassegrain is used for an astronomy course. "Funds for public viewing have not been provided." writes a correspondent. For an active public program in North Dakota, see Minot State College.

MINOT STATE COLLEGE OBSERVATORY. Minot, N.D. 58701. On campus. Public hours 9 p.m. to 2 a.m. Thur evenings, and by arrangement. Telescopes include a Celestron 16-inch, and two others of 6- and 5-inch aperture, plus a custom 6-inch refractor. There is a visual and ultraviolet spectrophotometer. All these are used mainly for educational purposes. This is a fine facility for visual observing, and the only one in the country to announce public hours extending past midnight.

Other observational telescopes:
Dickinson State College

Ohio

BURRELL MEMORIAL OBSERVATORY. Baldwin-Wallace College. 36 E. 5th Ave., Berea, Ohio 44017. About 14 mi. SW of Cleveland. Visits 7–8 p.m. one Fri per month, usually near the first-quarter moon, Sept–Mar. Exhibits of meteorites; antique instruments, especially timekeepers; and transparencies of celestial objects. The 13-inch refractor with its 4-inch finder is used mainly for teaching and observation. There is also a 3-inch transit. The director has various research interests, but no projects are currently under way on these instruments. The Observatory was built in 1940 with funds donated by the widow of Edward P. Burrell, chief engineer for many years with Warner and Swasey. Low elevation (720 feet) near both Cleveland and Lake Erie leads to rather bad conditions for observing.

CEDARVILLE COLLEGE OBSERVATORY. P.O. Box 601, Cedarville, Ohio 45314. On campus, 2 blocks W of SR 72 and 4 blocks N of US 42 and the SR 72 intersection. Viewing and lectures for visitors scheduled by request. The main telescope is a 16-inch Cassegrain of 56-inch focal length. There also is a 6.5-inch refractor of 96-inch focal length and auxiliary instrumentation consisting of two photoelectric photometers, a photopolarimeter system, and an astro-camera. Special studies are carried out in eclipsing variable stars and planetary atmosphere. The original Observatory had a 10-inch Newtonian with which studies on RX Herculis were pursued; the new center was dedicated in 1973.

CINCINNATI OBSERVATORY. University of Cincinnati, Ohio 45208. On Observatory Place. Visits 8:30 to 9:30 p.m. by *reservation only*. Regular tours about ten nights per month; reservations required. Phone: (513) 321-5186. The principal instrument, which is also the one used for visitors' viewing, is an 11-inch refractor that O. M. Mitchell bought in Munich in 1841. The Clarks worked it over in 1876; one would like to have heard the elder Alvan Clark's reflections on the occasion, since he had witnessed the importation of Harvard's "Great Refractor" at huge expense from Germany—and had discovered errors of figure with his own eye. The refurbishing of the Cincinnati telescope was a measure of the esteem enjoyed at that time by American telescope makers; the Clark firm, in fact, earned a reputation abroad before it did at home. The Observatory at one time owned a short-focus 4-inch Clark "comet seeker," and a Fauth-mounted transit with a 5½-inch Clark objective. Other equipment has also been used, including a 48-helical beam radio telescope, one of the earlier such experiments. Much work has been carried out here in dynamical astronomy, and Cincinnati is the Minor Planet Center of the International Astronomical Union.

MUELLER PLANETARIUM AND OBSERVATORY. Cleveland Museum of Natural History, Wade Oval, University Circle, Cleveland, Ohio 44106. Take University Circle exit off I-90 on Shoreway. Telescope viewing on clear Wed evenings Sept–April. Adult education courses in astronomy. Write for planetarium schedule. The fine Brashear 10½-inch refractor is supported by the Cleveland hallmark, a Warner and Swasey mounting. There is some auxiliary photoelectric equipment, but the research astronomer on the staff observes elsewhere. Museum exhibits and a bookstore complete this important center for public astronomy, which offers introductory astronomy instruction for most age groups.

PERKINS OBSERVATORY. Ohio Wesleyan and Ohio State University, Delaware, Ohio 43015. On U.S. 23, 2 miles S of Delaware, Ohio, or about 20 miles N of Columbus.

Write for schedule of Public Education Services. Open 2–3 p.m. Mon–Fri and on guest nights by advance registration only. Send self-addressed, stamped envelope to obtain free tickets. Observation is possible with the 32-inch reflecting telescope, one of the largest in the country that the public may look through. The 32-inch occupies a dome that formerly housed the 69-inch reflector that was moved to Flagstaff in 1961; subsequently the mirror was replaced with a 72-inch one of low-expansion glass, and this is now the largest instrument at the Lowell Observatory. Near the Observatory in Delaware is a 16/24 Schmidt used by faculty and graduate students. The very active Ohio State University Astronomy Department here possesses a number of sophisticated instruments and a large astronomical library. Specialized programs continue on late-type stars, planetary nebulae, stellar atmospheres, stellar rotation, solar spectroscopy, and laboratory astrophysics.

WARNER AND SWASEY OBSERVATORY. Case Western Reserve University, 1975 Taylor Road, East Cleveland, Ohio 44112. Write for yearly schedule of lectures and open evenings. Two consecutive evenings per month; telephone reservations required (Phone (216) 451-5624). Fri evening observations during July and Aug. A few astronomical exhibits. The Observatory owns a large Schmidt with 24-inch aperture and 36-inch mirror that is used for direct and objective prism photography 30 miles east of Cleveland, at its Nassau Station. A 36-inch reflector at the Observatory proper handles programs in spectroscopy, photoelectric photometry, and polarimetry. A 9½-inch refractor made in 1890 with optics by Brashear serves as a teaching and demonstration instrument. The Warner and Swasey Observatory was founded in 1920 with the support of Mr. W. R. Warner and Mr. A. Swasey who presented their own refractor to the Case School of Applied Science, and paid for the building to house it. The Schmidt was acquired in 1940 and the 36-inch in 1957. These facilities are used by the staff of the Department of Astronomy of Case Western Reserve University for research and instruction. An undergraduate degree in astronomy is offered as well as the Ph.D.

ASTRONOMICAL SOCIETY OF FOSTORIA. Fostoria, Ohio. An observatory is planned here to house the 12½-inch refractor made by John Brashear in 1896 which was in the Emerson McMillan Observatory until 1962. The McMillan Observatory is being demolished because of deterioration and light pollution on the Ohio State campus in Columbus. It is hoped that the Warner and Swasey mounting can be restored.

SWASEY OBSERVATORY. Denison University, Granville, Ohio 43023. Visits by appointment for class-size groups; open houses sometimes for community or University-connected persons. The 9-inch refractor on its Warner and Swasey mounting has a Brashear doublet objective optimized for visual wavelengths. There are visual, photographic, and photometric accessories. The Observatory also owns four portable telescopes, including a new 8-inch. From 150 to 200 students a year observe with the equipment, and physics majors use it for senior honors projects. Photometric observation of Algol with subsequent computer reduction of results has determined its period with state-of-the-art precision. The physics department has just completed a Czerny-Turner long path spectrograph to be used for molecular spectrography and is "confident of accomplishing significant work in this field." The Observatory was dedicated in 1910 with some of the best equipment available at the time; this included a transit instrument, a chronograph, and precision pendulum clocks.

SCHOONOVER OBSERVATORY. Jefferson St. in Schoonover Park, Lima, Ohio. Public nights announced in local newspapers; groups may arrange visits with the Lima

Astronomy Club. The 12½-inch Newtonian-Cassegrain has a mirror that is made from some pyrex left over when the mold for the 200-inch Hale telescope was filled by the Corning Glass works; there is a 4-inch refractor finder and a portable 4-inch refractor. The Observatory was built with funds left to the city by T. R. Schoonover, with the Astronomy Club, which meets there the 4th Wednesday each month, providing the telescope. Over 10,000 people have visited it.

RITTER ASTROPHYSICAL RESEARCH CENTER. The University of Toledo, 2801 Bancroft, Toledo, Ohio 43606. The telescope is opened for public observing at 8:30 and 9:30 on a Friday near the middle of the month. There is no charge, but reservations must be made. No group reservations are accepted. The Ritter Planetarium has shows Fri, 7:30 p.m. and Sun 2 p.m. Admission for adults is $1, students $.50 and children $.25. School classes can have special shows on Tue and Thurs mornings. Write for leaflet, or call (419) 537-2650. The Ritter 40-inch telescope, built in 1967 by Warner and Swasey, is the first telescope equipped with "zero-expansion" Cervit optics. It is presently the largest telescope in Ohio. The telescope can be operated as a Cassegrain or coudé reflector by use of a flippable secondary mirror. The Cassegrain spectrograph, built by Optics for Industry, can be used with an image tube. A high resolution echelle spectrograph is under construction. A 4-inch astrographic camera is mounted on the telescope along with two 6-inch refractors serving as guide telescopes. The Observatory also has a Gaertner oscilloscope comparator for measuring and tracing stellar spectra. The well-equipped darkroom facility includes a vacuum-chamber system for sensitizing photographic plates with hydrogen gas. Research with this facility is largely on variable stars such as flare and magnetic stars.

Other observational telescopes:
Denison University
Mt. Vernon Nazarene College
Otterbein College
Wittenberg University
Wright State University
Xavier University
Youngstown State University

Cincinnati Museum of Natural History
Supplementary Educational Center, Cleveland
Battelle P., Center of Science and Industry, Columbus
Ohio State University, Columbus
Dayton Museum of Natural History
Kent State University, Kent
Malabar High School, Mansfield
Sandusky High School
Warren City Planetarium
Youngstown State University, Youngstown

Larger planetariums:
Firestone High School, Akron
Lake Erie Junior Nature and Science Center, Bay Village
Stark County Historical Center, Canton

Oklahoma

UNIVERSITY OF OKLAHOMA OBSERVATORY. Dept. of Physics and Astronomy, 440 West Brooks, Room 131, Norman, Oklahoma 73069. Located at 1314 Asp in Norman. There is a program of two public nights per month during the semester. The 10-inch Newtonian with an 85-mm Zeiss astrograph mounted on it is used for teaching. Professor B. S. Whitney accumulated an archive of 20,000 plates for photographic photometry of eclipsing binaries and variables between 1942 and 1962. There is also an 8-inch Newtonian for demonstration purposes, and a 12-inch Cassegrain for photoelectric photometry will be installed in an off-campus observing station which is under construction. Theoretical work is carried on in stellar atmospheres, evolution of close

145

binaries, and cosmology. There is a BS in astrophysics and graduate students may work on astrophysical problems for a Ph.D. in physics. One of the darkest skies, with the best view of the milky way, that the compiler of this directory ever saw was in the Fountain-head State Park to the east of Norman; except for problems with wind, the plains states are much underrated as observatory sites.

Other observational telescopes:	Larger planetariums:
Central State College	Oklahoma Science and Arts Foundation,
Southwestern College	Oklahoma City

Oregon

PINE MOUNTAIN OBSERVATORY. Dept. of Physics, University of Oregon, Eugene, Ore. 97403. Located in central Ore. 50 miles E of the Cascade Mountains and 30 miles SE of Bend, Ore., at an elevation of 6,300 feet. Public hours Thurs through Sun, afternoons and evenings. Equipment consists of a 24-inch Cassegrain and another 15-inch reflector. The three astronomers on the staff have put this equipment to better use than have many researchers with access to larger apertures: E. G. Ebbighausen works with precision determinations of binary light curves; James Kemp developed a highly accurate device to measure circular polarization and was the first to show the presence of a large magnetic field for a white dwarf; and Ira Nolt devised a highly sensitive liquid-helium-cooled germanium bolometer for work in the medium and near infrared. The Observatory, which was built in 1967, is among the best-utilized of medium-sized facilities in the country. The whole region east of the Cascades provides numerous attractive areas with clear skies where amateurs on summer vacations might wish to set up camp and use their own equipment.

LEWIS AND CLARK OBSERVATORY. Portland, Ore. 97219. Not open to the public. The 10-inch Newtonian is altogether taken up with research (photoelectric light curves of eclipsing binaries).

Other observational telescopes:	Larger planetariums:
Portland Community College	Medford Senior High School, Medford
Mt. Hood Community College	Oregon Museum of Science and
Southern Oregon College	Industry, Portland
Treasure Valley Community College	

Pennsylvania

KEPLER OBSERVATORY. Temple University, Ambler Campus, Ambler, Pa. 19002. On Meeting House Rd. Public nights as announced; for information write Dr. Howard Poss, Physics Dept., Temple University, Philadelphia, Pa. 19122. The telescope is a 14-inch Celestron equipped for high-speed photoelectric photometry; it is used for educational purposes as well as photometry.

DICKINSON COLLEGE CELESTARIUM. Carlisle, Pa. 17013. Located in the Tome Science Bldg. on Louther St. between West and College St. Four public shows per semester, usually one each month from Oct to May. Write to arrange group visits. Many interesting exhibits that can probably be viewed during any class day. This is an unusually complete and intelligently planned center for astronomical education. The **ROSCOE O. BONISTEEL PLANETARIUM** now has a Goto-Venus model 5500 projector, an unusually fine "skyline," and seats specially designed for planetarium viewing. There is a Foucault pendulum, a Cavendish gravitational apparatus, two orreries, and a solar chronometer. A waiting-room-lounge with a 6-inch telescope, a slide projector, and numerous books and charts keep those waiting their turns at the telescopes happy and informed. An open deck, the best arrangement for public viewing, contains four 10-inch Criterion reflectors, and there are several other telescopes including a Questar, an 8-inch Tinsley Cassegrain, and an 8-inch Criterion. All these facilities are used in the Dickinson College astronomy courses.

EDINBORO ASTRONOMICAL OBSERVATORY. Dept. of Physics and Astronomy, Edinboro State College, Edinboro, Pa. 16444. On Campus. Two or three public nights per semester, as announced. This Observatory possesses some impressive photographic equipment, with both 16-inch and 8-inch Schmidt cameras as well as a 16-inch Celestron; a Williams cold camera permits long exposures without reciprocity failure. High-resolution spectroscopy of bright stars has been carried out with a Heath-Kit spectrophotometer. Degrees are offered in physics and science education.

RIDLEY OBSERVATORY AND PLANETARIUM. Ridley School District, Folsom, Pa. 19033. Astronomy facilities are primarily for educational use, research programs being conducted by faculty and students. Write for further information and programs to Nicholas Ignatuck, Jr., director, or call (215) 534-1900. The Observatory contains a 20-inch Newtonian reflector presented to the school by the Rittenhouse Astronomical Society of Philadelphia. The amateur-built instrument is one of the first telescopes of such size constructed by amateurs. The Observatory also contains a 10-inch Newtonian, an astrograph, an 8-inch richest-field reflector, a 6-inch f/8 Newtonian, a 4½-inch f/10 Newtonian and a 3-inch f/15 refractor. The 20-inch is equipped with a photometer with a 1P21 photocell and amplifier. Output is recorded on a strip chart recorder. The observatory is used year round by students from seventh to twelfth grade. The planetarium is used to supplement observational activities as well as for interdisciplinary demonstrations with other departments. Observational interests include astrophotography (stellar and diffuse objects including comets), photometry (long-period variable stars) and air pollution studies. Nearby areas suitable for use of amateur instruments include Ridley Lake, Ridley Creek State Park, and Chadds Peak.

STRAWBRIDGE OBSERVATORY. Haverford College, Haverford, Pa. 19041. The Observatory is on the campus and visits may be arranged for groups. At one time the Clarks, according to D. J. Warner, repolished an 8¼-inch refractor made by Henry Fitz (whose optics shop is reconstructed in the Smithsonian). Even with its figure corrected, the lens was not very good, so the college bought a 10-inch Clark which is still in use. The Astronomy Dept. carries out investigations in radio astronomy, cosmology, and stellar atmospheres and interiors.

KUTZTOWN STATE COLLEGE OBSERVATORY. Kutztown, Pa. 19530. Located in the Grim Science Bldg. Visits and showings may be arranged with the Observatory and its associated planetarium. The Observatory has a Tinsley 18-inch Cassegrain which is used for photoelectric photometry of eclipsing binaries. The well-equipped planetarium seats 103 and uses a Spitz A3P-R. An athletic field next to the planetarium may be used for amateur telescopes.

JOSEPH R. GRUNDY OBSERVATORY. Franklin and Marshall College. Lancaster, Pa. 17604. Baker Campus of the College, north end of Wilson Drive. First and third Fri of each month *if the sky is clear.* The 16-inch Boller and Chivens Cassegrain is used for photoelectric photometry of variable stars. A large roll-off roof observatory also houses the Scholl 11-inch refractor, with lens made by Alvan Clark and mounting by the German firm of Repsolds and Sons; it is used for teaching and general observation. Another instrument of historical interest is the 5-inch Brashear refractor, which may be the first made by that second-greatest early American optician (who also made the 30-inch at Allegheny Observatory and the 72-inch mirror for the Dominion Observatory in Victoria, B.C.). There are several other instruments. There has been an observatory here since 1884. The director writes: "Around the turn of the century the Hamilton Watch Company had a telegraph line to the Scholl Observatory. The astronomer in charge determined meridian transits of stars, obtained the local sidereal time and zone time, then sent this information by wire to the Hamilton Watch Company. At one time we were the standard for the watch company!" The College has a stimulating program enabling physics and math majors to prepare for further work in astronomy. The campus adjacent to the Observatory is suitable for amateur observation. Also on the campus, at the corner of College and Buchanan Avenues, is the **NORTH MUSEUM;** this includes a 41-foot dome planetarium with a Spitz AP4 projector, and free shows are offered Sat at 3 and Sun at 2 and 3. There are numerous natural history exhibits. Write for a current brochure.

BUCKNELL UNIVERSITY OBSERVATORY. Lewisburg, Pa. 17837. On US 15 just south of Lewisburg, at the rear of Bucknell University Stadium. Visits scheduled by request. The 10-inch Clark refractor has been in use since 1887; it has a 12-foot focal length. Studies in the internal structure of stars are pursued at Bucknell, and there are courses in astronomy and astrophysics. To the west are a number of campgrounds that could be used for amateur viewing.

DREXEL UNIVERSITY OBSERVATORY. 32nd and Chestnut Sts., Philadelphia, Pa. 19104 No regular public hours. The 10-inch Celestron and the 12-inch Newtonian are used for teaching and for student observation, with a few student research projects being carried out.

FELS PLANETARIUM. The Franklin Institute, 20th and the Parkway, Philadelphia, Pa. 19103. Open Tue–Sat 10–5; Sun 12–5. Planetarium open Fri 8 p.m. Admission: $2, plus $.50 for planetarium. Three floors of physical science exhibits. Shop and bookstore. Telescopic observation when possible during museum hours and after Fri planetarium show. A Zeiss 10-inch refractor and a 24-inch reflector are used only for public observation. This is a major center for popular education in astronomy, with some of the best telescopes for public observation in the Northeast.

FLOWER AND COOK OBSERVATORY. University of Pennsylvania, Philadelphia, Pa. 19174. Not open to the public. The main site of the Observatory is in Malvern, Pennsylvania, 18 miles west of Philadelphia. Current major telescopic equipment includes a 28-inch Cassegrain reflector, a 27-inch Cassegrain reflector under construction and a unique 15-inch horizontal refractor with siderostat. An 8-inch Clark refractor, a 4-inch astrographic camera, and a 3-inch broken transit are mounted at the campus station of the Observatory. Semiautomatic two-channel radiometric and polarimetric measuring systems are available for each telescope. Founded in 1895. Research in galactic clusters, binary stars, radiometry, polarimetry, and development of observational techniques and instrumentation. This Observatory, which is a consolidation of the older Flower Observatory and Cook Observatory, has a rich and interesting history, and the astronomers and astrophysicists associated with it carry out a wide variety of observational, experimental, and theoretical research programs. Some observational research carried out in the past: detection of the inner corona without a solar eclipse by electronic scanning techniques, pioneer development of pulse-counting techniques in stellar radiometry, development of automatic two-channel systems for radiometry and polarimetry, photographic spectroscopy of unusual stars, precise measurement and analysis of the light curves of many eclipsing binaries, radiometric studies of galactic clusters, Fourier analysis of the structure of gaseous nebulae. Jointly with the University of Canterbury in New Zealand and the University of Florida, the Flower and Cook Observatory operates the Mount John University Observatory on the South Island of New Zealand. Located there and belonging to Flower and Cook are a 10-inch astrographic camera and a 24-inch Cassegrain reflector. One of John A. Brashear's largest lenses, of 18-inch clear aperture, was lent to Percival Lowell in 1894 for his early observations of Mars at Flagstaff, Arizona; subsequently it was installed at the Flower Observatory in a mounting by Warner and Swasey and is now at the New Zealand station.

BUHL PLANETARIUM AND INSTITUTE OF POPULAR SCIENCE. Allegheny Square (north side), Pittsburgh, Pa. 15212. Routes 8 and 28; any major route to downtown Pittsburgh. Open every day 1–5 p.m. and *also* Wed–Sat 7–10 p.m. and Sat 10:45–5 p.m.; Sun 1–10 p.m. Admission: Sky Drama and Inst.: adult $1.50, children $.70. Inst. only: adult $1.25, children $.65. Sky shows at 2:15 and 8:15 p.m. when open; extra shows Sat 11:15 a.m. and Sun 4:15 p.m. The Zeiss model II projector provides the Sky Shows. There is also the only 10-inch siderostat (a large aperture telescope with a mirror that rotates to follow the object viewed) available for public observation. Weather and staff permitting, solar observation is possible before and after the afternoon Sky Show, and viewing of the moon and planets before and after evening shows. There are very extensive natural history exhibits of all kinds, most of which have some bearing on astronomy. These include a Foucault Pendulum, which demonstrates the rotation of the earth; "Your weight on other worlds" exhibit by the Toledo Scale Company; murals of astronomical subjects; a diorama of the Barringer Meteor Crater; exhibits on the history of astronomy, and much more. It is a first-rate center for public education in astronomy.

CARNEGIE-MELLON UNIVERSITY. Dept. of Physics, Schenley Park, Pittsburgh, Pa. 15213. Although the University does not at this time own an observatory, and offers no public nights, it does own several good portable instruments for use in astronomy instruction. Furthermore, it owns the 11-inch refractor built for Andrew Carnegie by Brashear, whose early career kept him in this city working as a millwright. Simeon A. Friedberg, Chairman of the Physics Dept., writes: "Negotiations have been under way for some time with local amateur groups, and we hope eventually that the telescope can have a permanent home, where it could be used by students and public alike."

SPROUL OBSERVATORY. Swarthmore College. Swarthmore, Pa. 19081. Off Rt. 320. Open second Tue each month during the academic year, 7–8:30; later in March, Apr, May. Individual visitors (no groups) and no children seventh grade or under. Some exhibits. Occasional visits arranged for recognized amateur groups by writing to the Director. The large 61 cm refractor with its 10.93-meter focal length was made by Brashear and mounted by Warner and Swasey in 1911. Such instruments lend themselves to astrometry because of their stable optical performance; the Sproul refractor is used in photographic astrometry of nearby stars, (the faintest of which cannot be studied in the Philadelphia urban environment). A special feature is a systematic search for stars with invisible companions which could be dim dwarfs, black holes, or large planets. A small oscillatory component in a star's proper motion reveals these. There is also a 6-inch Clark refractor. At one time the Observatory lent Percival Lowell a wide-field 9-inch Brashear refractor to aid in his search for Pluto.

VILLANOVA UNIVERSITY OBSERVATORY. Villanova, Pa. 19085. Ten miles W of Philadelphia on US 30. Planetarium shows and observatory tours on request for small groups. Scheduled public shows once a month during the academic year. Instrumentation includes a 15-inch Cassegrain, a 10-inch Schmidt-Cassegrain, a 6-inch refractor, and an 18-foot radio dish (2,000 MHz) plus a 30 MHz radio telescope for monitoring outbursts from Jupiter. The University offers a BS in astronomy; research programs include photoelectric photometry of Be stars and magnesium photometry of bright field stars.

PENNSYLVANIA STATE UNIVERSITY RADIO ASTRONOMY OBSERVATORY. Dept. of Astronomy, 102 Whitmore Laboratory, University Park, Pa. 16802. Equipment consists of a half-dozen small dishes for observation of the radio sun, plus a 10-meter paraboloid for the same purpose. Other small dishes and various types of arrays carry out sky mapping, solar, and galactic studies. A particular specialty is the observation of sudden bursts of solar activity.

PENNSYLVANIA STATE UNIVERSITY OBSERVATORY. Department of Astronomy, University Park, Pa. 16802. Located about 30 miles NW of Penn State, on Rattlesnake Mountain, near Black Moshannon Reservoir. Write for information about visits. Penn State has just installed a 60-inch Astro-Mechanics Cassegrain with an f/2 primary mirror made of Tensalloy aluminum. (For other instruments of this type, see Capilla Peak Complex in Arizona). The idea is to get more aperture than possible with a smaller conventional mirror, recognizing that there may be a sacrifice in resolution. The steps in the making of the mirror are interesting, since they are somewhat unconventional; the metal blank was machined to the approximate figure, then coated with nickel to provide a hard surface for optical finishing. The surface was then aluminized, as with a glass mirror. The telescope rides in an English yoke mounting. A two-channel photometer is being fitted to the telescope, and will collect data to be stored in a desk computer. A laboratory spectrograph is being modified to be used on the telescope, either photographically or with a multichannel analyzer. The aluminum mirror may also serve as a radio telescope for short-millimeter wavelengths. Penn State also has a 24-inch Cassegrain at the same site, which has been used to observe quasars. Black Moshannon State Park provides a dark-sky site that amateurs might wish to use.

Other observational telescopes:
Butler County Community College
California State College (Pennsylvania)
Clarion State College
Indiana University of Pennsylvania
Juniata College
Mansfield State College
St. Vincent College
Thiel College
West Chester State College
Westminster College

Larger planetariums:
Louis E. Dieruff High School, Allentown
Freedom High School, Bethlehem
Cedar Cliff High School, Camp Hill
Charleroi Area High School, Charleroi
Clarion State College, Clarion
Lehigh Valley Astronomical Society,
   Coopersburg
Gettysburg College, Gettysburg
William Penn Memorial Museum,
   Harrisburg

150

Hatboro-Horsham Senior High School, Horsham

Indiana University of Pennsylvania, Indiana

Pequea Valley Intermediate School, Kinzers

Greater Latrobe Senior High School, Latrobe

Lock Haven State College, Lock Haven

Burrell High School, Lower Burrell

Penn Hills Senior High School, Pittsburgh

Science Motivation Center, Pottsville

Reading School District, Reading

West Chester State College, West Chester

York Suburban High School, York

Puerto Rico

ARECIBO OBSERVATORY. National Radio and Ionospheric Center, P.O. Box 995, Arecibo, Puerto Rico 00612. Operated by Cornell University for the National Science Foundation. Located 10 miles south of the town of Arecibo, near Esperanza. Visiting hours Sun afternoon 2–4:30. Overlook of the antenna system and exhibits of pictures. This is the largest single-element radio antenna dish in the world for use as a radio telescope. The spherical reflecting dish is 1,000 feet in diameter, and it is set into a natural basin in rugged mountain terrain. The basin represents a familiar feature in karst geological areas; it is a collapsed limestone cave system. The location is a good one for escaping radio interference. Originally the surface of the reflector was wire mesh, but in late 1974 a new surface of perforated aluminum panels was completed. This added enormously to the efficiency of the antenna and extended the spectrum of the radio waves that it could be used to receive by a factor of ten. The perforations are there to let in light that permits vegetation beneath the dish to grow and stabilize the dirt surface;

Aerial view into the 1,000-foot Arecibo antenna. Towers support the receiving elements at the focus; note the curved track to move the focal point. (Cornell University Photograph)

wind-loads are not much of a problem, as they are in large fully steerable antennas. Another part of the renovation included installing of an extremely powerful S-band transmitter for radar mapping of planetary surfaces. This should make it possible to distinguish altitudes of 100 meters' difference on the surface of Venus; it may well turn out that this is the best "view" of that planet's surface that we ever get prior to close-up radar mapping from a satellite in orbit around Venus. An ingenious system of altitude-sensors on the cable-supported central focal assembly raise and lower it by activating screws in the cable supports. This keeps the receiving (or transmitting) aerials at the focal surface. The antenna can be steered to any point within 20 degrees of the zenith. This means that certain objects can be tracked for as long as three hours, despite the diurnal motion of the earth's surface; so this is not merely a transit instrument (see Green Bank: the 300-foot telescope). Receiving antennas can be displaced along a track north and south of the meridian, covering a zone of the sky that usually includes the sun, moon, and planets. The sensitivity of the instrument is far more important than these directional limitations, considering the amount of work left to do in all parts of the sky and in all branches of radio astronomy. The Arecibo antenna is, for example, currently adding enormously to the knowledge about and identification of organic molecules in galactic space. The Observatory has just recently put out an excellent pamphlet that covers both its equipment and its research programs in more detail than this account.

Other observational telescopes:
University of Puerto Rico

Rhode Island

LADD OBSERVATORY. Brown University, 210 Doyle Ave., Providence, R.I. 02912. Corner of Hope and Doyle, about 1 mi N of campus. Open Wed Sept–May 8 p.m. and June–Aug 8:30. Lectures Oct–May; at least three programs per month, usually two speakers and a film. Exhibits include a transparency room, various nineteenth-century instruments, and a 3-inch transit in the campus library that once gave a time standard for Providence. The 12-inch refractor, made by John Brashear, and the 3-inch transit in the Observatory are used for teaching and public viewing. Professor Francine Jackson writes: "Completed in 1891, Ladd Observatory is named for Herbert Warren Ladd, governor of R.I. at that time. It is located on the highest point of the city, once a dump appropriately named Tin Can Alley." Professor Jackson goes on to explain that light pollution makes the Observatory useful mainly for teaching and observing, and expresses concern that economy moves at Brown may result in its being closed down. Teachers and amateurs who find this a valuable public service in this large metropolitan area might find it useful to see if public or private measures can be taken to help keep it open. Amateurs desiring darker skies might look into the Seagrave Observatory off Rt. 116 in N. Scituate. It should be remembered that excellent views of the moon and planets (and on occasion some of the brighter nebulae and clusters) are easy to obtain even in conditions that make deep-sky photography or research quite difficult. David Targan writes: "Staff members are conducting several programs of instruction with the main instrument and others, and teaching continues to be the main function of this observatory. When the observatory was founded, in 1891, some of the most prominent scientists of the 19th century were participants in the dedicatory exercise. Among them were Professor E. Pickering of Harvard, Professor C. Hastings of Yale, J. Brashear of Pittsburgh, and Alvan Clark of Cambridge. It was Hastings who designed the lens of the main instrument. Brashear, considered by some to be among the top three telescope makers in America, made the 12-inch lens, following Hastings' instructions. At the time of the dedication, the observatory was equipped with a very fine high-precision filar micrometer, a stellar spectroscope, and a 3-inch transit telescope. The 3-inch telescope was used for many years to observe stars crossing the meridian. These observations provided the data necessary to compute the exact time. A regular program of transit observations and time keeping was started in 1893. This program was discontinued in 1916, and reestablished in 1919. The transit instrument is still functional today, although the time keeping system is in a state of disrepair. Dr. Charles H. Smiley came to Brown in 1930, and became the director of Ladd Observatory in 1938. Professor Smiley became well known for his knowledge of Mayan astronomy, and his solar eclipse expeditions. The nature of the research that was undertaken at the Observatory under his direction was primarily mathematical (such as the determination of orbits). Smiley was also responsible for building a 12-inch Schwarzschild Camera, one of two such instruments in the world. The Observatory has recently acquired more modern instrumentation. In 1971, four 3½-inch Questar telescopes and four cameras were purchased, including a Schmidt camera. The Observatory building itself is a two-story structure. The upper story consists of an outside deck and the observing room. The outside deck is used presently for telescopic observations with Questars and other small, teaching instruments, and for general outdoors viewing, such as observing the constellations. The observing room houses the main instrument, the 12-inch Brashear refractor. The first floor has a small lecture room, a clock room, a transparency room, and a small room for equipment storage. Adjacent to the main structure of the Observatory is the transit room, which contains the transit telescope and some displays of antique astronomical instrumentation."

ROGER WILLIAMS PARK MUSEUM PLANETARIUM.   Providence, R.I. 02905. Broad St. or Elmwood Ave. Open Mon–Thur 8:30–4; Fri 8:30–3; Sat 9–4:30; Sun 2–5. Planetarium shows (subject to change) Sat 2:30 and 3:30; Sun 3 and 4. Bookings for schools on weekdays by appointment. Free admission. Exhibits on anthropology and natural sciences, and a sales desk for books and scientific items. The planetarium has a 24-foot dome and a Spitz A-2 projector with special-effects devices. The museum is located in the middle of an extensive park maintained by the city of Providence.

Larger planetariums:
Space Science Lab, Middletown Schools

South Carolina

MELTON MEMORIAL OBSERVATORY.   University of South Carolina, Columbia, S.C. 29208. On the USC campus at the corner of Green and Bull Sts. Visiting hours Oct–April, Mon 8:30–10:30 p.m. May–Sept, Mon 9:30–11:30 p.m. Subject to change. Additional hours arranged for groups, often prior to Mon public hours, Oct–April. Some exhibits of photographs. The main instrument is a 40-cm Cassegrain/Newtonian on a Warner and Swasey mount, with a Brashear mirror made in 1880. There is a 4-inch Unitron guide scope attached. There are other small telescopes, and an aerial camera converted for astrophotography. The camera is located 12 miles outside Columbia. Work can be done in photometry and special studies. The Observatory is primarily a teaching and public service center, with modest amounts of undergraduate research. Amateurs may use their own instruments at the R. G. Bell Camp (USC) observation station, in Sesqui Park, and at Fort Jackson.

Other observational telescopes:
College of Charleston
Converse College
Furman University
Presbyterian College

Larger planetariums:
Charleston Museum
Bob Jones University, Greenville

South Dakota

BROOKHAVEN SOLAR NEUTRINO OBSERVATORY. Lead, S.D. Not open to the public. Deep underground in the Homestake Gold Mine there is a 100,000 gallon cylindrical tank containing 610 tons of perchloroethylene cleaning fluid. Neutrinos from the sun are supposed to react with the chlorine in this compound to produce radioactive Argon-37. Periodically the tank is purged with helium and the radioactive argon is counted. Results so far have not indicated as many neutrinos as expected from theories of the solar energy generation process. Raymond Davis, Jr., who devised the project, also writes: "There is also at our location a set of Cerenkov particle detector systems designed to observe the neutrino pulse from a collapsing star (super-nova). A second detector system is also located in the Barberton Limestone Mine, Pittsburgh Plate Glass Co., Ohio."

Other observational telescopes:
Sioux Falls College
University of South Dakota

Tennessee

KING COLLEGE OBSERVATORY. Bristol, Tenn. 37620. Public nights as announced. The 12½-inch reflector can be used with a photoelectric photometer; there is also a 10-inch mobile telescope on a trailer that can be taken to eclipses and to dark-sky sites. Under construction is a 10-inch sidereostat. Research projects on the variability of Ap stars and binary systems are carried out by physics majors; photoelectric detection of small light variations in ellipsoidal variables is another specialty.

CLARENCE T. JONES OBSERVATORY. University of Tennessee at Chattanooga, Chattanooga, Tenn. 37402. On Brainerd Rd. Public viewing every Fri Sept–May at 8 p.m. Barnard Astronomical Society meets there 8 p.m. the second Thur of each month. The 20-inch Cassegrain was at one time one of the largest amateur instruments in the country; it and a small planetarium are now used for public programs and in connection with astronomy courses.

UNIVERSITY OF TENNESSEE. Dept. of Physics and Astronomy, Knoxville, Tenn. 37916. Public nights Mon and Fri evenings on the Physics Bldg. roof, where there is a 4½-inch mounted Bausch and Lomb refractor. The Dept. also has an 8-inch Celestron with a portable tripod. Both telescopes are used for teaching and public service. An introductory astronomy course and advanced undergraduate astrophysics course are offered. Some research is being carried out by Dr. Kenneth Fox in planetary atmospheres.

RICE OBSERVATORY. Cumberland College, Lebanon, Tenn. 37087. This is a well-equipped small college observatory, with a 12-inch reflector and a Questar, as well as the historically and optically interesting 7-inch refractor made by Alvan Clark. In early 1975 the Observatory reported: "not open at this time—needs some repair."

LABORATORY OF ATMOSPHERIC AND OPTICAL PHYSICS. Southwestern at Memphis, 2000 North Parkway, Memphis, Tenn. 38112. On the campus. Tours throughout the year to students and other people interested in the facilities. Public viewing sessions about four times during the school year. Telescopes include a 31-inch reflector that can be used at $f/2$ or $f/4.5$; a 24-inch long-focus ($f/12$) reflector, a 6-inch refractor, and a 7-inch Questar; auxiliary equipment consists of a high-resolution grating spectrometer for work in visible and ultraviolet, infrared prism and grating spectrometers, infrared radiometers, and so on. On display (and operating) is a diffraction grating ruling engine of the Rowland type. There was a mobile observatory built in 1960 for eclipse expeditions and field studies. A specialty here is spectroscopic and radiometric studies of the solar atmosphere in the infrared. A coelostat on the roof of the Physics Tower brings solar radiation through trap doors in the floors to most of the laboratories in the building. "We have a history," writes J. H. Taylor, "of approximately two decades now of studying total solar eclipses. On these expeditions, as well as others, we make every attempt to involve as many of our undergraduates as possible . . . . We were a member of the 1973 American Solar Eclipse Expedition to Africa."

ARTHUR J. DYER OBSERVATORY, Vanderbilt University, Nashville, Tenn. 37240. About 7.5 miles south of downtown Nashville, out 12th Ave. which turns into Granny White Pike. Then left on Oman Dr. Monthly visitors' programs; write for schedule. Groups wishing to visit the Dyer Observatory should write to the director; a limited number can be admitted (adults only) if Observatory duties permit. Children may come to regular visits with their parents, but it is better to ask about special times for children. Interested persons may join the Barnard Astronomical Society, which includes telescope-making, lectures, and observing in its programs; write to the Observatory secretary. The booklet, *Astronomy at Vanderbilt*, may be the most attractive, informative, and complete publication by any observatory. The original Observatory was named for E. E. Barnard, one of Vanderbilt's earliest students and perhaps the last great self-taught astronomer. The man responsible for the new Observatory was Carl Seyfert, whose name attaches to the peculiar galaxies with bright centers and emission lines in their spectra (this fact is for some reason not mentioned in the booklet). The Seyfert telescope is a unique 24-inch reflector engineered so that it can be used as Cassegrain, Newtonian, or a sort of variant on the Schmidt, employing a wide-field corrector near the plate. Equipment includes an objective prism and photoelectric and spectrographic devices. On visitors' nights viewing is possible through the 24-inch and also a 12-inch; illustrated talks and films complete the program. Research programs are chiefly directed towards galactic structure; projects beyond the Observatory's equipment capability have been scheduled at Kitt Peak and Cerro Tololo. Darkrooms, laboratories, and an extensive astronomical library support the research programs. A smaller observatory on the top floor of the Physics and Astronomy Building of Stevenson Center for Natural Sciences (on the Vanderbilt campus) houses a 6-inch Cook refractor that was originally in the Barnard Observatory. As *Astronomy at Vanderbilt* describes it, few other private universities in the country offer such a comprehensive program in astronomy, from the most sophisticated research to a sensitive and public-spirited accommodation of public interest and educational need.

UNIVERSITY OF TENNESSEE SPACE INSTITUTE. Tullahoma, Tenn. 37388. A 12½-inch Newtonian reflector will be installed in a dome here, and it is expected that it will be accessible to the public and to amateur astronomers.

Other observational telescopes:	Larger planetariums:
Cleveland State Community College	Lambuth College, Jackson
University of the South	Nature and Interpretive Center, Bays Mountain Park, Kingsport
	Sudekum P., Nashville Children's Museum

Texas

BEE CAVES OBSERVATORY. Astronomy Department, University of Texas at Austin, Austin, Tex. 78712. Located on a 1,000-foot hill, 10 miles SW of the center of Austin. Facilities include the Calvert 12-inch Newtonian and Couder twin 8-inch refractors; there is also an instrument shop. The Austin Amateur Astronomers have a 12.5-inch Cassegrain on the same site. This relatively dark site (an ex-NIKE base) serves many of the observing needs of the large undergraduate astronomy program at U. T. Austin and the active local amateur groups.

MILLIMETER WAVE OBSERVATORY. Electrical Engineering Research Laboratory, University of Texas at Austin, Rt. 4, P.O. Box 189, Austin, Tex. 78757. Located on Mt. Locke. The 4.85-m paraboloid observes all available objects in the millimeter-wave band, including the sun, moon, planets, galaxies, and extragalactic sources. The fully steerable dish is made of invar, and is figured to a highly precise tolerance of 0.1 mm surface accuracy. It is useful to wavelengths less than 1 mm. Harvard, Bell Telephone Labs, and NASA operate this jointly with U. T.

PAN AMERICAN UNIVERSITY OBSERVATORY AND PLANETARIUM. Edinburg, Tex. 78539. Until very recently this University had an extremely active program in astronomical education and in public service. Equipment in 1974 included a 17-inch reflector on the campus and a 16-inch reflector at a dark-sky site 15 miles away. There was also a large NASA camera system and an 8-inch Schmidt camera. A recent letter states "we will no longer be offering public nights at our observatory." Pan American University has had an excellent undergraduate program in astronomical education, and has been the location of a good deal of equipment; so presumably the whole program has not been closed down.

HARVARD RADIO ASTRONOMY STATION. Fort Davis, Tex. 79734. A radio telescope of 26-meter diameter is used for observations of galactic and extragalactic objects and, at times of sunspot maximum, solar flares. A smaller radio telescope of 9.5 meters diameter, and other smaller antennas, are used for continuous observations of solar activity. The Station came into operation in 1956.

MARGARET ROOT BROWN OBSERVATORY AND BURKE BAKER PLANETARIUM. Houston Museum of Natural Science, 5800 Caroline, Houston, Tex. 77025. West corner of Hermann Park. Museum open (free admission) Tue–Sat 9–5 and Sun–Mon 12–5 plus Fri–Sat eves. 7:30–9:00. Some free lectures; write for schedule. Planetarium shows May 31–Sept 2, Mon–Fri at 2 and 3, Sat–Sun at 2, 3 and 4. Fri–Sat eves. at 8 p.m. Sept 6–June 1; Wed and Fri at 4 plus same Fri–Sun schedule as summers. Admis-

Other types of radio antennas for radio astronomy. These are used by the University of Texas; for an extensive description of radio astronomy with such antennas, read about Clark Lake, California. (Photo: courtesy Astronomy Department, the University of Texas)

sion: Adult $1.25; children under 12 $.50; no children under 5 admitted. Reservations at (713) 526-4273. School groups may arrange other times for a small fee or none at all. The Spitz projector is augmented by more than fifty auxiliary projectors making possible all manner of special effects. Solar viewing in the museum is via a 3.6-inch telescope linked to projection screens and a TV camera. The 16-inch reflector in the Observatory also can be used for group viewing with a TV link, and is open regularly each month as well as being available for inspection during the day. Special observing sessions may also be requested by calling (713) 526-4273. This complex is another example of one of the excellent centers that have come into existence in this country for the purpose of providing popular education in astronomy and the other sciences. The rest of the museum provides exhibits of all kinds, some of which bear on astronomy. There is a program of Sat morning classes in astronomy. Write for current schedules. Surrounding Hermann Park may be used for amateur observation.

RICE UNIVERSITY. Dept. of Space Physics and Astronomy. Houston, Tex. 77001. All instruments are used for teaching or research, and include some optical telescopes of 12-inch and smaller aperture, plus gamma-ray telescopes 6-inch and smaller, with which the Crab Nebula pulsar's gamma rays were first observed, and with which nuclear gamma rays from galactic and extragalactic sources were discovered. In addition to the gamma-ray studies, also called "nuclear astronomy," the department specializes in optical and ultraviolet photometry.

UNIVERSITY OF TEXAS RADIO ASTRONOMY OBSERVATORY. Dept. of Astronomy, University of Texas, Austin, Tex. 78712. Located near Marfa, Tex. A six-element two-by-two-mile 380 MHz synthesis interferometer is conducting a survey of discrete radio sources, obtaining arc-second positions, structure and flux densities for about 50,000 objects, together with information on radio source variability. Two-element interferometers at 30, 22.2, 20.0, 16.7 and 10.0 MHz are used to monitor the sporadic decametric radiation from Jupiter. A 3-station spaced receiver network and 22.2 and 16.7 polarization analysers provide information on Stokes' parameters and coherence properties of the Jovian radiation.

Other observational telescopes:
Alvin Junior College
Eastfield College
McClennan Community College
Richland College
San Antonio Union College
Texas A & I University
TCU
University of Houston

Larger planetariums:
Abilene Public Schools
South Park Independent School District, Beaumont
Dallas Health and Science Museum
St. Mark's School, Dallas
El Paso Independent School District
San Antonio College
Tyler Junior College, Tyler
Richfield High School, Waco

Utah

UTAH STATE AERONOMY/ASTRONOMY OBSERVATORY. University Hill, Logan, Utah 84322. Public tours by appointment only. This Observatory uses a very specialized assembly of equipment for making observations of infrared radiation. There is a wide-field interferometer-spectrometer cooled by liquid nitrogen and liquid helium. Supporting this apparatus are photometric and radiometric collectors, a tracking mount, and a computerized data-handling system. Observations are made of hydrogen-alpha radiation that has been doppler-shifted into the infrared, O_2 transitions, and time-resolved OH. Amateur astronomers passing through the area can find beautiful Forest Service campsites with dark skies along U.S. 89 between Logan and Bear Lake.

BRIGHAM YOUNG UNIVERSITY OBSERVATORY. Provo, Utah 84602. No public nights. A 24-inch Cassegrain is used with a photometer, a spectrograph, and a photoelectric scanner to study variable stars and eclipsing binaries. BS, MS, and PhD degrees are offered in physics and astronomy. The University also owns the Summer Hayes Planetarium.

UNIVERSITY OF UTAH OBSERVATORY. Physics Dept., Salt Lake City, Utah, 84112. Observatory on South Physics Bldg., near intersection of 1st South and Butler St., near 13 East. Open evenings by appointment; call receptionist at (801) 581-6905. The principal instrument, used for teaching, is a 24-inch Ealing Cassegrain. Dark-sky sites may be found at Little Mountain and at Grantsville, which is near the Wasatch National Forest.

Other observational telescopes:
Dixie College
Hansen Planetarium
Southern Utah State College
Utah State University
Utah Technical College at Provo

Larger Planetariums:
Weber State College, Ogden

Vermont

BENNINGTON COLLEGE. Bennington, Vt. 05201. No regularly scheduled use for the 10-inch Celestron, which is used for teaching.

CASTLETON STATE COLLEGE. Castleton, Vt. 05735. Open for the public occasionally. The Unitron 6-inch refractor is equipped with filters, three cameras, and two spectroscopes. It is mainly for student use.

MIDDLEBURY COLLEGE OBSERVATORY. Middlebury, Vt. 05753. Public hours from time to time, announced locally. The 10-inch Celestron here is only used for teaching, but the Middlebury Astrophysics Project specializes in X-ray astronomy.

THE SPRINGFIELD TELESCOPE MAKERS. Springfield, Vt. 05156. Every Aug the most celebrated convention of amateur telescope makers in the country assembles in Springfield, Vt. For advance details, consult *Sky and Telescope.* In connection with the meeting, the observatory equipment at **STELLAFANE** is made available for viewing, including the Porter turret telescope. A project is currently under way to restore the nearby Hartness turret telescope, a unique contraption somewhat like a military tank buried in the ground. A museum will be developed to occupy the underground room beneath the turret. Information was not supplied about the visiting hours at Stellafane during other times in the year; but the annual meetings in early August have been going on for more than forty years. (See reports in *Sky and Telescope* for October and November, 1975.)

Virginia

KEEBLE OBSERVATORY. Randolph-Macon College, E. Patrick St., Ashland, Va. 23005. Visits, guided tours, and viewing can be arranged. Some public nights. The 12-inch Cassegrain and a Questar are mostly used for student projects. At one time the College owned a Clark-figured refractor. There are displays in the Science Building lobby. Amateurs may set up telescopes in an area immediately adjacent to the Observatory.

VIRGINIA POLYTECHNIC INSTITUTE. Department of Physics, Blacksburg, Va. 24061. Public nights 2nd and 4th clear Fri each month. A variety of visual instruments make it possible to demonstrate many important optical configurations to students. These range from a 14-inch Celestron to a 3-inch "Copernik" Maksutov, with two 6-inch Criterions, two 4¼-inch Edmund Newtonians, and an 8-inch Celestron in between.

CHESAPEAKE PLANETARIUM AND OBSERVATORY. 300 Cedar Rd., Chesapeake, Va. 23320. Take route 168 South to Great Bridge. Free observation each clear Thurs night throughout the year. Open other nights coinciding with events of celestial importance. The telescope is a 14-inch Celestron modified for closed circuit television. In addition to public astronomy the telescope and equipment carry out solar studies and record meteor showers. The planetarium seats 130 under a 30-foot dome. The center was the first of its kind built by the public school system in Virginia. A unique facility for those in the Tidewater area.

LANGLEY RESEARCH CENTER. NASA, Hampton, Va. 23365. Take exit 8C from I-64 onto Magruder Blvd.; then Shepherd Blvd. to Gate 4 of Langley Air Force Base. This involves reversing your direction on I-64 if you are coming east, from Richmond, by covering half the cloverleaf at exit 8. Open 8:30–4:30, except Sun noon to 4:30. The Visitor Center has extensive displays on space exploration, moon rocks, movies, interplanetary probes, etc.

WALLOPS STATION. Radio Astronomy Branch, Goddard Space Flight Center, Greenbelt, Md. 20771. Located at Wallops, Va. The 18.3-meter paraboloid is used to observe the radio continuum and the lunar occultation of radio sources; it also participates in experiments in Very Long Baseline Interferometry.

KLINK OBSERVATORY. Virginia Military Institute, Lexington, Va. 24450. Open during the school year by appointment. A 16-inch amateur-built Newtonian is used for teaching, public observation, and some variable star work. Available to groups of children and adults.

WILLIAM AND MARY OBSERVATORY. Dept. of Physics, College of William and Mary, Williamsburg, Va. 23185. On the campus. The 10-inch Criterion reflector and a Questar are currently used for informal observing in astronomy and cosmology classes, and also for the public nights which are announced in local newspapers. A larger telescope will soon be added to the Observatory.

Other observational telescopes:

Hampden-Sydney College
Hollins College
Madison College
Virginia Commonwealth University
Virginia Technical University

Washington

GOLDENDALE OBSERVATORY. Owned by the City of Goldendale and leased to Goldendale Observatory Corporation. Goldendale, Wash. 98620. Open irregularly, depending on the availability of volunteer guides. Currently open Sat evenings. Inquire locally. Available to college and high school groups. Equipment, all for visual observing, includes a 24-inch Cassegrain, a 10-inch Cassegrain, a 6-inch Newtonian, a 6-inch refractor, and two 3-inch refractors; 8- and 10-inch Newtonians will be added.

JEWETT OBSERVATORY. Washington State University, Pullman, Wash. 99163. "Free public hours on an irregular basis." Although the respondent to a questionnaire did not list anything under "Instruments of historical interest," the 12-inch refractor at this observatory is an Alvan Clark product, the date of which is not yet determined. This may be of some interest to residents of the Northwest, since not many of these long-focal-length telescopes of the nineteenth century migrated in that direction.

BATTELLE OBSERVATORY (also known as Rattlesnake Mountain Observatory). Battelle-Northwest Labs., P.O. Box 999, Battelle Blvd., Richland, Wash. 99352. 20 miles west of Richland. Visits in spring, summer, and fall by appointment. Call (509) 946-2383 or (509) 942-7301 about tours. A 31-inch reflecting telescope is used for stellar and planetary polarimetry; a 30-foot radio telescope takes measurements in the radio spectrum (millimeter wavelength) and monitors the intensity of radio sources. The Observatory is located within the ERDA Hanford Reservation on a treeless ridge in the southeastern Washington desert, near the confluence of the Columbia, Snake, and Yakima rivers, at an elevation of 3,560 feet. Average yearly rainfall is less than 6.3 inches. The first full year of operation for the optical telescope was 1972, and the radio dish began operation in 1975. The Observatory was first organized in 1967 around an airglow and auroral observing program; this is still a major activity. Around Yakima to the north are several campgrounds, and in the eastern parts of the Cascades there are numerous beautiful camping areas that enjoy many clear nights.

PACIFIC SCIENCE CENTER. 200 2nd St. N., Seattle, Wash. 98109. Open summers 11–8; winter weekdays 9–5, weekends 11–8. Admission: adults $2, students $1, children $.50, pre-school free, whole family $5; senior citizens $.50. Not exactly a planetarium, but the Center does have a domed "Spacearium" with shows that often have astronomical subjects; there are five buildings full of exhibits.

UNIVERSITY OF WASHINGTON. Astronomy Dept. Seattle, Wash. 98195. The campus observatory is open to the public every clear Thur night during the academic year. Closed summers. About four times a year the **MANASTASH OBSERVATORY**

at Ellensburg, Wash., is open, as announced in local papers. On the campus the visitor may be shown slide shows about astronomy, and there are some exhibits of photographs and of instruments, such as transits and computers. A 6-inch refractor made by Brashear serves well for public observing. The 30-inch reflector at Manastash is used for photography and photoelectric photometry. There are numerous National Forest campgrounds throughout the Cascade Mountains. There, and near Ellensburg, there are also good places for independent camping. Maps are available from Forest Service Offices. Areas near Ellensburg enjoy the dry climate in the rain shadow of the mountains; low population density and less tourist travel than some areas should make this an attractive refuge for the amateur astronomer.

Other observational telescopes:
Big Bend Community College
Edmonds Community College
Evergreen State College
Seattle Pacific College
Spokane Falls Community College
University of Puget Sound
Western Washington State University
Whatcom Community College

Larger planetariums:
Western Washington College,
  Bellingham
Eastern Washington College, Cheney
Yakima Valley College, Yakima

West Virginia

CONCORD COLLEGE OBSERVATORY. Athens, W. Va. 24712. On campus, 1 mile east of Rt. 20. Open for anyone two nights per week during the academic year. There is a 6-inch Unitron refractor; a 10-inch reflector not permanently mounted; five portable Edmund 4¼-inch reflectors; and a Questar. These are used in connection with an astronomy course. It is hoped that by 1978 the Unitron will be in a new dome atop a new library wing.

Larger planetariums:
Charleston Children's Museum
Wood County P., Parkersburg
Brooke High School, Wellsburg

Wisconsin

THOMPSON OBSERVATORY. Beloit College, Beloit, Wisc. 53511. Located on the campus in Chamberlin Hall, at the corner of Emerson and Pleasant Sts. Public viewing sessions about ten times per year; scheduled irregularly. The Observatory is now equipped with a 22-inch Celestron and nine 10-inch Celestrons, the largest collection of instruments of such aperture for student use. A specialty is the development of spectrographic instruments. Students interested in astronomy take a bachelor's degree in physics. Until 1967 the principal telescope was a 9½-inch Alvan Clark refractor. It had been installed in 1882 on a Warner and Swasey mounting. Dave Garroway bought the telescope from Beloit, and the mounting went to the Smithsonian; when Garroway decided to go back to television, he put the refractor back up for sale.

CASEY OBSERVATORY. University of Wisconsin at Eau Claire, Wisc. 54701. No public hours. A 14-inch Celestron, a 12-inch Newtonian at a remote station, and a 1-meter

Roland solar spectrograph make up the equipment; special studies in astrometric positions of minor planets and comets and some sort of galactic studies are carried out.

PINE BLUFF OBSERVATORY. University of Wisconsin, Madison, Wisc. 53706. Near the village of Pine Bluff, 15 miles west of Madison. Annual open house on a Sun in May. The main instrument here is a 36-inch reflector of the type known as Dall-Kirkham; this has an ellipsoidal primary mirror and a spherical secondary that sends the light back through a perforation as in an ordinary Cassegrain. The optics were figured in the optical shop at Yerkes, and the mount is by Boller and Chivens. There is also a 16-inch reflector. Research is carried out in spectrophotometry, continuing the tradition of photoelectric work established at Washburn Observatory (q.v.). Astronomers connected with this observatory carry out programs elsewhere as well; two are involved with NASA's plans for a large orbiting telescope, and auxiliary equipment developed here has been used with the 4-meter Mayall reflector at Kitt Peak. The Observatory is located on a hilltop on a 53-acre site, and amateur astronomers may obtain permission to use their telescopes here.

WASHBURN OBSERVATORY. University of Wisconsin, Madison, Wisc. 53706. On Observatory Drive in Madison. Open first and third Wed of the month, *if clear*, 7:30 p.m. in winter and fall, and 9 p.m. spring and summer. The 15-inch refractor is used for teaching and for student viewing. The actual aperture of the instrument is 15.6-inches, and the reason for this is that the Clarks were instructed to make a lens larger than Harvard's. Although the Washburn telescope was already outranked in size by two other Clark products (the 18½-inch Dearborn and the 26-inch Naval refractors), this must have stirred memories for Alvan Clark. The first real telescope he had a look at was Harvard's Great Refractor, made by the German firm of Merz and Mahler, and his own efforts were actually stimulated by slight defects that he could discern in its lens. The Observatory itself was a gift of Governor Cadwallader C. Washburn, and on its completion in 1879 it was the westernmost U.S. observatory of any importance. Much pioneering work in photoelectric photometry was carried out here. One sometimes reads about such equipment that it can "detect the flame of a candle" at such-and-such a distance. According to an account by Joel Stebbins, he and his colleagues finally decided that they had better make good on such assertions, so they set up an actual candle on the other side of the lake from the Observatory, and did in fact find its light easily detectable on their photometer, even without using a telescope. There is also an Astronomy Dept. planetarium which offers free monthly shows. For more information about astronomy at the University, see the entry above on Pine Bluff Observatory.

BUCKSTAFF OBSERVATORY. University of Wisconsin at Oshkosh, Wisc. 54901. "Used for student work only due to lack of personnel." The 16-inch Cassegrain and the 3-inch refractor have been used for many observations of variable stars and sunspots. The Observatory was built in the early 1920s by Ralph Buckstaff and given to the University about 1960. Mr. Buckstaff was for many years active and prominent in the AAVSO, and sent in sunspot records for decades.

Other observational telescopes:	Larger planetariums:
Carthage College	Wisconsin State University, Eau Claire
University of Wisconsin at Stevens Point	University of Wisconsin, Kenosha
and at Whitewater	University of Wisconsin, Madison
Wisconsin State University	University of Wisconsin, Menasha

Hamilton High School, Milwaukee
Madison High School, Milwaukee
Gifford Junior High School, Racine

Wisconsin State University, Stevens
  Point
Wausau West High School, Wausau
Casper P., Casper

Wyoming

WYOMING INFRARED OBSERVATORY. Dept. of Physics and Astronomy, University of Wyoming, Laramie, Wyo. 82071. Located on Jelm Mountain, Centennial, Wyo. This telescope, scheduled to go into operation in 1977, will put this observatory at the forefront of infrared observing. It will have a 90-inch mirror with a Cassegrain focus, and will no doubt incorporate the very latest technological remedies for preventing interference from ambient heat sources. The primary mirror will probably be made of Cervit and figured to an f/2.1 curve; it will ride in an English yoke mounting. Using the sort of technology developed at M.I.T.'s Wallace Observatory and being adapted to the Lick 120-inch, the telescope will be computer-controlled. The site on Jelm Mountain at 9,656 feet of elevation provides the thin, dry air best for infrared observations. Thirty percent of the observing time will be reserved for guest observers. The University of Wyoming also owns Celestrons of 5- and 8-inch aperture; inquire about viewing hours on campus.

SOURCES AND BIBLIOGRAPHY

The publications mentioned in this paragraph are here because I consulted them in compiling a file of names and addresses of observatories. Three were also especially helpful in fleshing out or interpreting questionnaires returned by observatories. These included *Sky and Telescope* (Sky Publishing Corporation, Cambridge, Mass.), nearly 400 of whose issues I searched; the *Bulletin of the American Astronomical Society* (American Institute of Physics, N.Y.) whose Observatory Reports provided current technical information on the more active research centers; and Deborah Jean Warner's book, *Alvan Clark and Sons, Artists in Optics (United States National Museum Bulletin 274,* Washington D.C., 1968), which indicated locations of telescopes of great historical interest and furnished facts about them. I trust that a general acknowledgment to all three of these excellent publications will suffice in place of a plethora of footnotes. Actual quotation is attributed as it occurs; I could not help using the stylistically felicitous Warner book. Other sources of observatory names, in approximate order of helpfulness, include the Navy's *American Ephemeris and Nautical Almanac;* the National Academy of Science's *1974 List of Radio and Radar Observatories;* Norman Sperling's looseleaf *A Catalog of North American Planetariums* (1971); Claire Inch Moyer's *Silver Domes* (Denver, 1955) which provided many useful facts as well as addresses; the National Academy of Science report: *Astronomy and Astrophysics for the 1970's;* lists provided by the Celestron Company and by Ash Domes; various academic directories; and tips from many individual persons as well as scattered references, in many publications, to the existence of establishments. The Smithsonian Astrophysical Observatory generously provided computer print-outs of their address lists, which served as a most valuable cross-check.

Other books consulted frequently, mostly for historical facts:

Berry, Arthur. *A Short History of Astronomy.* London, 1898; republished New York: Dover, 1961.

Brashear, John A. *The Autobiography of a Man who Loved the Stars.* New York: American Society of Mechanical Engineers, 1924.

Collins, A. Frederick. *The Greatest Eye in the World.* New York: Appleton-Century, 1942.

Fassero, James S. *Photographic Giants of Palomar.* Los Angeles: Westernlore Press, 1947.

Jones, Bessie Zaban. *Lighthouse of the Skies.* Washington, D.C.: Smithsonian Publication 4612, 1965.

Jones, Bessie Zaban and Boyd, Lyle Gifford. *The Harvard College Observatory.* Cambridge: Harvard, 1971.

Kuiper, Gerard P. and Middlehurst, Barbara M. *Telescopes.* Chicago: University of Chicago, 1960.

Miczaika, G. R. and Sinton, William M. *Tools of the Astronomer.* Cambridge: Harvard, 1961.

Pendray, G. Edward. *Men, Mirrors, and Stars.* New York: Funk and Wagnalls, 1935.

Struve, Otto and Zebergs, Velta. *Astronomy of the 20th Century.* New York: Macmillan, 1962.

United States National Museum Bulletin 228: Holcomb, Fitz, and Peate: Three 19th Century American Telescope Makers. Washington: 1962.

GENERAL BOOKS ABOUT ASTRONOMY FOR FURTHER READING

(Listed in order of increasing difficulty or technicality).

Stars, by Herbert S. Zim and Robert H. Baker; Golden Press (A Golden Nature Guide).
The Sky Observer's Guide, by R. Newton Mayall and Margaret W. Mayall; Golden Press (A Golden Nature Guide).
Astronomy, by Iain Nicholson; Bantam (A Knowledge Through Color Book).
Dynamic Astronomy, by Robert T. Dixon; Prentice-Hall.
Introduction to Astronomy, by Laurence W. Fredrick and Robert H. Baker; Van Nostrand Reinhold.
Introductory Astronomy and Astrophysics, by Elske v.P. Smith and Kenneth C. Jacobs; W. B. Saunders.

INDEX

Numbers in parentheses give aperture in inches of larger telescopes at each location, or in feet when identified as radio telescopes. Small-aperture solar instruments are simply called "solar." Only the proper names of installations are listed here; see the catalog for listing by state and for institutional affiliation. All names below designate observatories unless otherwise indicated.

An asterisk (*) denotes instruments of historical interest, usually refractors.

171